PERGAMON INTERNATIONAL LIBRARY
of Science, Technology, Engineering and Social Studies
The 1000-volume original paperback library in aid of education, industrial training and the enjoyment of leisure
Publisher: Robert Maxwell, M.C.

Elements of Analytical Dynamics

INTERNATIONAL SERIES IN
PURE AND APPLIED MATHEMATICS
Volume 105

Some Other Titles in This Series

CHIRGWIN & PLUMPTON
A Course of Mathematics for Engineers and Scientists (6 volumes)

HAMMER
Elementary Matrix Algebra for Psychologists and Social Scientists

HILL
Principles of Dynamics

KURATOWSKI
Introduction to Calculus, 2nd edition

MELNYK
Principles of Applied Statistics

ROMAN
Some Modern Mathematics for Physicists and other Outsiders (2 volumes)

SMIRNOV
A Course of Higher Mathematics (5 volumes)

For Full lists of titles in Pure and Applied Mathematics see p. 183

ELEMENTS

OF

ANALYTICAL DYNAMICS

by

Rudolph Kurth

PERGAMON PRESS

OXFORD · NEW YORK · TORONTO
SYDNEY · PARIS · FRANKFURT

U. K.	Pergamon Press Ltd., Headington Hill Hall, Oxford OX3 0BW, England
U. S. A.	Pergamon Press Inc., Maxwell House, Fairview Park, Elmsford, New York 10523, U.S.A.
CANADA	Pergamon of Canada, Ltd., P.O. Box 9600, Don Mills M3C 2T9, Ontario, Canada
AUSTRALIA	Pergamon Press (Aust.) Pty. Ltd., 19a Boundary Street, Rushcutters Bay, N.S.W. 2011, Australia
FRANCE	Pergamon Press SARL, 24 rue des Ecoles, 75240 Paris, Cedex 05, France
WEST GERMANY	Pergamon Press GmbH, 6242, Kronberg-Taunus, Pferdstrasse 1, Frankfurt-am-Main, West Germany

First edition 1976

Library of Congress Catalog No. 75-21681

Printed in Great Britain by A. Wheaton & Company, Exeter

ISBN 0 08 019848 1 Hardcase
ISBN 0 08 019864 3 Student Flexi

CONTENTS

PREFACE

The present book has grown from notes of lectures given by the author at the Georgia Institute of Technology. Those lectures were delivered to first-year graduate students of Mathematics with little background in Theoretical Mechanics, and formed a part of a course in Topological Dynamics. This origin of the book may explain its scope.

Once again I extend my best thanks to my respected colleagues Professors W. Bennewitz, C. Ho, and A. Pal for their most valuable constructive criticisms, and to Mrs. S. Schaefer for her patient typing.

April 1975 R. Kurth

CHAPTER I

THE EQUATIONS OF MOTION

§ 1. Space and Time

1.1. Dynamics deals with the motions of material bodies. Since motion is defined as the change of position in the course of time, first an attempt will be made to clarify the notions of space and time. The terms "Space" and "Time" (with capital "S" and "T") will mean the "real", "actually experienced" space and time -- in contradistinction from conceptual constructions made in analogy to Space and Time. Such constructions will be denoted by the words "space" and "time" written with small initials.

1.2. The notions of Space and Time are taken as primitive; that is, they are left undefined. Basic properties of Space and Time are described by the following postulates:

I. Space is a connected analytic manifold.

II. Time is a 1-dimensional analytic manifold which is homeomorphic to the 1-dimensional Euclidean number space, E^1.

The definitions of the topological concepts used in these postulates will be recalled in the following subsections, and a justification of the postulates will be attempted at the end of this section.

1.3. The basic notions of set theory are supposed to be familiar; in particular those of: set (family), element (member); (infinitely) countable set; subset; union and intersection of sets, of a family of sets; disjoint sets; the Cartesian product of (finitely many) sets, the Cartesian power of a set; equivalence relation and equivalence class.

1.4. The n-dimensional Euclidean number space E^n is defined as the set of all n-tuples

$$x = \begin{pmatrix} x^1 \\ x^2 \\ \cdot \\ \cdot \\ \cdot \\ x^n \end{pmatrix}$$

of real numbers $x^1, x^2, \ldots, x^n$, endowed with the inner product

$$x^* y \underset{\text{Def.}}{=} \sum_{i=1}^{n} x^i y^i$$

for any two vectors (i.e. elements of E^n) x and y. An asterisk (*) will always denote the transpose of a matrix.

To any vector $x \varepsilon E^n$ there corresponds a non-negative real number $|x|$ defined by

$$|x| \underset{\text{Def.}}{=} +\sqrt{x^* x} \quad ;$$

it is called the norm of the vector x. Now let x be any given "point" (any element) of E^n, and ε be any positive real. Then the set

$$U(x;\varepsilon) \underset{\text{Def.}}{=} \{ y \varepsilon E^n \,\big|\, |y - x| < \varepsilon \}$$

is called a spherical neighbourhood (or, more precisely, the ε-neighbourhood) of x. A subset G of E^n is said to be open if for any point x of G there is a spherical neighbourhood of x which lies entirely in G. For convenience the empty set is also taken as an open subset of E^n.

We observe that:

(i) The union of any family of open subsets of E^n is open.

(ii) The intersection of any two open sets is open.

1.5. These notions and observations yield the pattern for the following definitions:

Let Γ be any family of sets such that:

(0) The empty set is a member of Γ.

(i) The union of any subfamily of Γ is a member of Γ.

(ii) The intersection of any two members of Γ is a member of Γ.

Then the family Γ is called a topology, and the set $X \underset{\text{Def.}}{=} \bigcup \Gamma$ is said to be the space of the topology Γ. The space X, endowed with the topology Γ, is called a topological space and denoted, for example, by (X,Γ). The elements of the space X are also said to be its points, and the members of the topology Γ are called the open (sub)sets of the space X(relative to the topology Γ), or the open subsets of the topological space (X,Γ).

Any open set which contains a given point p of X is said to be an open neighbourhood of that point p. A topological space is called a Hausdorff space if any two of its points have disjoint open neighbourhoods.

A topological space which is the union of two disjoint non-empty open subsets is said to be disconnected. A topological space which is not disconnected is called connected.

1.6. The notion of a function (or mapping) and the various notations used for functions are supposed to be familiar. So are, in particular, the notions of: domain, range; image, pre-image; mapping one-to-one, into, onto; inverse mapping; the restriction of a mapping to a (non-empty) subset of the domain; the composition of mappings.

Now let (X,Γ) and (Y,Δ) be topological spaces, $f : X \to Y$ be a mapping from X into Y, and p be any given point of X. The function f is said to be continuous at the point p if for any open neighbourhood V of $q \underset{\text{Def.}}{=} f(p)$ (a subset of Y) there is some open neighbourhood U of p (a subset of X) such that

$$f(U) \underset{\text{Def.}}{=} \{y \in Y \mid y = f(x) \text{ for some } x \in U\} \subseteq V.$$

The function f is said to be continuous in a subset A of X if it is continuous at every point of A.

If the mapping $f : X \to Y$ is continuous, and B is any open subset of Y, then $f^{-1}(B) \underset{\text{Def.}}{=} \{x \in X \mid f(x) \in B\}$ is an open subset of X, by the definition of continuity.

Let the mapping $f : X \to Y$ be one-to-one and onto, and both f and f^{-1} be continuous: then f is called a homeomorphism. If f is a homeomorphism, so is f^{-1}. If, for any two given topological spaces, there is a homeomorphism mapping the one onto the other, the spaces are said to be homeomorphic to each other. The relation "is homeomorphic to" is an equivalence relation: it is reflexive, symmetric, and transitive.

The homeomorphic image of an open set (i.e., the image of an open set under a homeomorphism) is an open set, by the definitions of "open set" and "homeomorphism."

1.7. Let (X,Γ) be a given topological space, and $\tilde{X}$ be any non-empty subset of X. Define, for any $G \in \Gamma$,

$$\tilde{G} = \tilde{X} \cap G,$$

and

$$\tilde{\Gamma} = \{\tilde{G} \mid G \in \Gamma\}.$$

Then, $\tilde{\Gamma}$ is a topology, and $\tilde{X}$ is its space. The topological space $(\tilde{X},\tilde{\Gamma})$ is said to be a subspace of (X,Γ). The sets $\tilde{G}$ are called open in $\tilde{X}$, or open in the ("relative") topology $\tilde{\Gamma}$.

If the mapping $f : X \to Y$ is continuous in X, then the restriction of f to any non-empty subset $\tilde{X}$ of X is continuous in $\tilde{X}$ with respect to the relative topology.

1.8. A subfamily B of a topology Γ is called a basis of Γ if every member of Γ is a union of members of B.

Thus any topology is a basis of itself. The family

$$\{U(x;\varepsilon) \mid \varepsilon \text{ is a positive real, } x \in E^n\}$$

is a basis of the n-dimensional Euclidean number space E^n or, rather, of the topology of E^n; and the family

$$\{U(x;\varepsilon) \mid \varepsilon \text{ is a positive rational number, } x \in E^n \text{ has rational coordinates}\}$$

is a countable basis of (the topology of) E^n.

1.9. Let (X,Γ) be a connected Hausdorff space whose topology Γ has a countable basis B. Suppose that for any point p of X there exists a member U_p of Γ which contains p and is mapped by a homeomorphism f onto an open subset $\underline{U}_p$ of the n-dimensional Euclidean number space E^n. Then (X,Γ) will be called an n-dimensional topological manifold, and the natural number n is called the dimension of (X,Γ). It is uniquely determined by the manifold (i.e., the manifold has exactly one dimension).

The homeomorphism f is said to be a system of local coordinates in the neighbourhood U_p, and the inverse homeomorphism, $f^{-1}: \underline{U}_p \xrightarrow{\text{onto}} U_p$, is called a local parametrization of the manifold (in the neighbourhood U_p). Any neighbourhood U_p on which local coordinates are defined will be called a coordinate neighbourhood.

1.10. Let U_p and U_q be two intersecting coordinate neighbourhoods on the n-dimensional topological manifold (X,Γ), with systems of local coordinates $f_p : U_p \xrightarrow{\text{onto}} \underline{U}_p$ and $f_q : U_q \xrightarrow{\text{onto}} \underline{U}_q$. Then the open subsets $f_p(U_p \cap U_q)$ and $f_q(U_p \cap U_q)$ of E^n are homeomorphic. The corresponding homeomorphisms between them, i.e., the mappings $f_q f_p^{-1}$ and $f_p f_q^{-1}$ restricted to the appropriate domains, are n-vector functions with continuous components.

If, moreover, these components have, for each pair of intersecting neighbourhoods, continuous derivatives of all orders (for short, if they are of differentiability class C^∞), then (X,Γ) (or, simply, X) is said to be an analytic manifold. Thus, if $x \to \underline{x}$ and $x \to \underline{\bar{x}}$ are any two of the local coordinate systems by means of which the analytic manifold has been defined, then both the coordinate transformation $\underline{x} \to \underline{\bar{x}}$ and the inverse coordinate transformation $\underline{\bar{x}} \to \underline{x}$ are of differentiability class C^∞.

Note that, for a given coordinate neighbourhood, the coordinate transformations of differentiability class C^∞ form a group, the composition of the mappings being the group operation. (The basic concepts of group theory are supposed to be familiar; in particular those of group, subgroup, and homomorphism.)

Any member of that group will be called an admissible transformation of local coordinates, and any system of local coordinates, obtained by an admissible coordinate transformation from that system of local coordinates by means of which the analytic manifold has been defined, will be called an admissible system of local coordinates.

1.11. Now all the topological concepts used in the Postulates I and II of §1.2 have been explained (viz. the concepts "n-dimensional analytic manifold," "connected," and "homeomorphic"). In short, those assumptions postulate that in Space and Time local coordinates can be introduced in such a way that any two overlapping systems of local coordinates are transformed into each other by means of real functions of differentiability class C^∞. In Time, moreover, always one single "global" system of coordinates can be used.

1.12. We now ask how the Assumptions I and II about Space and Time can be justified. Our answer consists of the following points:

(i) Physical experiences suggest that in Space and Time there are neighbourhoods of "Here" and "Now" in which local coordinates can be introduced.

(ii) A strong natural inclination of the human mind causes us to suppose that Space and Time are, in some sense, "homogeneous." We make the sense of "homogeneous" precise by postulating that systems of local coordinates with the same numbers of dimensions can be introduced everywhere in Space, and at any moment in Time.

(iii) If, for example, Space were disconnected, it would be hard to image that there could be any kind of physical interaction between (physical) "events" in "our" component of Space and those in any other component: it appears implausible that we could have any knowledge about it, even about its very existence - for us it is non-existing. Consequently we postulate that Space and, similarly, Time are connected. Thus, by (ii), they are topological manifolds.

(iv) Again experience suggests that, in overlapping neighbourhoods with local systems of coordinates, these systems can be chosen in such a way that in the intersection of those neighbourhoods the coordinate transformations are described by means of analytic, even of linear functions. Hence, by the "homogeneity principle," we postulate that Space and Time are connected analytic manifolds.

(v) The numbers of dimensions of these manifolds are 3 and 1. The 1-dimensionality of Time is essential in the theory of the mechanical phenomena observed in the real world, whereas the 3-dimensionality of Space appears in the theory often as accidental. Consequently, we postulate that Time is an 1-dimensional, but leave the number of dimensions of Space unspecified.

(vi) There are only two kinds of connected analytic 1-dimensional manifolds: they are homeomorphic either to the real line, or to the unit circle. By an extrapolation of experience, the human mind is inclined towards the belief that Time is linearly ordered globally, and not only locally; that is, for any two events which are not simultaneous, it makes sense to assert that the one was "earlier than" the other. Therefore we rule out the possibility of a circle and assume that Time is homeomorphic to the real line.

Notes and References

§ 1.2. The term "postulate" is used to indicate a hypothetical "first principle" which idealizes experience. The postulates introduced (together with those given later) suffice for the development of the theory; philosophically, however, they are insufficient. This can already be seen from the fact that, in some fashion, they re-introduce Newton's notions of absolute space and time. For a further discussion of Time, see Kurth (1964a); and for a historical survey over the problems of Space, see Jammer (1969).

§ 1.3. Standard references to set theory see Hausdorff (1914), Kamke (1928), Fraenkel (1966), Bourbaki (1968).

§§ 1.5 - 1.9. For the basic notions of "general" (or "set-theoretical") topology see, for example, Hausdorff (1914), Pontrjagin (1957), Kelley (1955), Bourbaki (1966), Schubert (1968). For the uniqueness of dimension see, for example, Munkres (1966), Hurewicz - Wallman (1948).

§§ 1.10 - 1.11. For basic notions of differential topology see, for example, Pontrjagin (1958), Milnor (1965), Munkres (1966). A proof of the theorem about 1-dimensional connected analytic manifold is given in Milnor (1965). For group theory see, e.g., Pontrjagin (1957).

§ 1.12. A more detailed discussion of the topological structure of Time is given in Kurth (1973a).

§2. Dynamical System of Particles

2.1. To "Space and "Time" (denoted by S and T) we now add "particle" as an undefined primitive notion. Postulate:

III. To any given particle there correspond a positive real number, called the mass of the particle, and a continuous mapping $g:T \to S$, called the motion of the particle.

For any "time", "instant", or "moment" t, that is, for any $t \in T$, the value g(t) of the function g is called the position of the particle at the time t. The n-tuple $\underline{x} \in E^n$ in a given system of local coordinates $x \to \underline{x}$ which corresponds to a position $x \in S$ of the particle will be called its position vector (relative to the chosen system of coordinates in S), and the real number $\underline{t}$ which corresponds to the instant $t \in T$ will be called the time (coordinate)(relative to the chosen system of global coordinates $t \to \underline{t}$ in T). The set of all time coordinates $\underline{t}$ will be denoted by $\underline{T}$.

2.2. A set of p particles (p being a natural number) will be said to be a system of particles if the positions of any two particles are distinct at all times, excepting at most countably many instants. In other words, there are, by definition, at most countably many collisions in a system of particles. (Collisions are discussed in Celestial Mechanics: this is the motivation for admitting, in the definition of a system of particles, the possibility of collisions.)

2.3. For the description of such a system the concept of configuration space can be useful. It is defined by means of the following auxiliary notions:

Let (X_1, Γ_1), (X_2, Γ_2) ,...,(X_q, Γ_q) be q given topological spaces, and B be the family

$$\{G_1 \times G_2 \times \cdots \times G_q | G_1 \varepsilon \Gamma_1, G_2 \varepsilon \Gamma_2 ,\cdots, G_q \varepsilon \Gamma_q\}$$

of subsets of $X_1 \times X_2 \times \cdots \times X_n$. Then the family Γ of all unions of members of B is a topology of the space $X \underset{(Def.)}{=} X_1 \times X_2 \times \cdots \times X_q$, and B is a basis of Γ. The topological space (X, Γ) is called the topological product of the topological spaces (X_1, Γ_1), (X_2, Γ_2), $\cdots$,(X_q, Γ_q). If all these spaces are identical, their topological product is said to be the qth topological power of any of them. Thus, for example, the n-dimensional Euclidean number space E^n, thought of as a topological space, is the nth topological power of the 1-dimensional Euclidean number space E^1.

The configuration space of a system of p particles is defined as the pth Cartesian power of the Space S. The ordered p-tuples of positions of the p particles are called configurations. They are points of the configuration space. A configuration will be said to be collision-free if any two of its positions are distinct. An instant of which the configuration of a system of particles is collision-free, will also be called collision-free.

An indirect argument, using the continuity of motion, shows:

To each collision-free moment $\overset{o}{t}$ of a system of particles there corresponds, in Time, a largest connected open neighbourhood T' of $\overset{o}{t}$ such that every instant t of T' is collision-free. The neighbourhood T' is uniquely determined by the motions of the particles and the instant $\overset{o}{t}$.

It will be called <u>the collision-free neighbourhood</u> of $\overset{\circ}{t}$ or, without reference to $\overset{\circ}{t}$, <u>a collision-free Time interval</u>.

2.4. Now suppose that:

(i) The Space S is homeomorphic to the n-dimensional Euclidean number space E^n;

(ii) The set of all coordinate systems $S \underset{\text{onto}}{\rightarrow} E^n$ includes a subset of distinguished coordinate systems, called systems of <u>rectangular Cartesian coordinates</u>.

The term "system of rectangular Cartesian coordinates" is taken as primitive.

(iii) For any two systems of rectangular Cartesian coordinates in S, $x \to \underline{x}$ and $x \to \underline{\bar{x}}$,

$$\underline{\bar{x}} = \underline{A}(t)\underline{x} + \underline{b}(t), \qquad t \in T;$$

here $\underline{A}(t)$ denotes an orthogonal $(n \times n)$-matrix and $\underline{b}(t)$ denotes an n-vector, both depending on the systems of coordinates considered and the instant t.

For a system of p particles with masses $\mu_1,\ldots,\mu_p$ and position vectors $\underline{x}_1,\ldots,\underline{x}_p$ (with respect to a given system of rectangular Cartesian coordinates), let

$$\underline{c} \underset{\text{Def.}}{=} \frac{\sum_{i=1}^{p} \mu_i \underline{x}_i}{\sum_{i=1}^{p} \mu_i} .$$

This equation is invariant under the transformations of coordinates $\underline{x} \to \underline{\bar{x}}$ considered in the third assumption above. Therefore the point c of S which corresponds to the vector $\underline{c}$ is independent of the system of rectangular Cartesian coordinates chosen. It is called the <u>centre</u>

of masses of the system of particles. In the following, only those systems of rectangular Cartesian coordinates will be considered whose origins are assigned to the centre of masses of the system of particles.

Finally suppose that:

(iv) There are a system of rectangular Cartesian coordinates $x \to \underline{x}$ in S, a global system of coordinates $t \to \underline{t}$ in T, and a given real function ϕ which is defined and has a continuous derivative for all positive real arguments, of the following kind: in any collision-free interval of Time, the position vectors $\underline{x}_1, \underline{x}_2, \ldots, \underline{x}_p$ of the particles are functions of the Time coordinate $\underline{t}$ which satisfy the "equations of motion"

$$\frac{d^2\underline{x}_i}{d\underline{t}^2} = \sum_{\substack{j=1,\ldots,p, \\ j \neq i}} \mu_j \phi(|\underline{x}_j - \underline{x}_i|) \frac{\underline{x}_j - \underline{x}_i}{|\underline{x}_j - \underline{x}_i|}, \quad i=1,\ldots,p.$$

The function ϕ is determined by the physical nature of the particles of the system. It is called the law of force of the system, and the system of particles itself will be called a free dynamical system of particles.

2.5. Throughout this book it will tacitly be assumed that any real-valued function of real variables discussed has continuous derivatives of all orders in its domain. In most, but not all, cases this assumption can be dispensed with. It has been made because the necessary differentiability conditions are not uniform throughout the theory and it appears of relatively little importance to list the precise "minimal" differentiability class in every individual case. Instead, the above uniform differentiability hypothesis has been adopted; but, once more, it could often be relaxed.

2.6. A straight-forward verification shows that the equations of motion of a free dynamical system of particles can be written in the form

$$\mu_i \frac{d^2\underline{x}_i}{d\underline{t}^2} = -\frac{\partial}{\partial\underline{x}_i} U(\underline{x}_1,\ldots,\underline{x}_n), \quad i=1,2,\ldots,p,$$

where

$$U(\underline{x}_1,\ldots,\underline{x}_n) \underset{\text{Def.}}{=} -\sum_{1 \leq i < j \leq p} \mu_i\mu_j \phi(|\underline{x}_i - \underline{x}_j|)$$

and, for $\rho > 0$,

$$\Phi(\rho) \underset{\text{Def.}}{=} \int \phi(\rho)d\rho \ .$$

The symbol $\frac{\partial}{\partial\underline{x}_i} U$ denotes the gradient of the function U with respect to $\underline{x}_i$, i.e., the n-dimensional vector function whose components are the partial derivatives of U with respect to the components of $\underline{x}_i$.

The equations are to be understood in the following way: first, differentiate the function U with respect to the independent variables $\underline{x}_i$; then substitute, for these independent variables $\underline{x}_i$, the motions $\underline{t} \to \underline{x}_i(\underline{t})$; and finally carry out, on the left-hand sides, the required differentiations with respect to $\underline{t}$: then both sides of the equation are equal for all admitted values of t. Thus the meanings of the symbol "$\underline{x}_i$:" is ambiguous: In "$\partial/\partial\underline{x}_i$" it means an independent variable, and at the other places of the formula it means the function $\underline{t} \to \underline{x}_i(\underline{t})$. Since confusion appears unlikely, similar convenient notations will be applied later also to the Lagrange and Hamilton equations.

2.7. The equations of motion can be given an even more compact form: Define (n p)-vectors X and U_X by

$$X = \begin{pmatrix} \underline{x}_1 \\ \underline{x}_2 \\ \vdots \\ \underline{x}_p \end{pmatrix}, \qquad U_X = \begin{pmatrix} \partial U/\partial\underline{x}_1 \\ \partial U/\partial\underline{x}_2 \\ \vdots \\ \partial U/\partial\underline{x}_p \end{pmatrix}.$$

so that the right-hand sides are partitions of the (np)-tuples X and U_X; let I_n be (n×n) unit matrix, and define the (np×np)-matrix M by

$$M = \begin{pmatrix} \mu_1 I_n & & & \\ & \mu_2 I_n & & \\ & & \ddots & \\ & & & \mu_p I_n \end{pmatrix},$$

all terms outside the main diagonal being 0. Then the equation of motion can be written as

$$M \frac{d^2X}{d\underline{t}^2} = -U_X(X)$$

<u>2.8</u>. This equation suggests a generalization of the concept of a free dynamical system of particles:

Suppose that, for some given systems of rectangular Cartesian coordinates $x \to \underline{x}$ in S and Time coordinates $t \to \underline{t}$, the position vectors $\underline{x}_1, \underline{x}_2, \ldots, \underline{x}_p$ of the particles are functions of the time variable $\underline{t}$ which, in any collision-free Time interval, satisfy the "<u>equation of motion</u>"

$$M \frac{d^2X}{d\underline{t}^2} = -U_X(X,\underline{t}).$$

The meaning of the symbols M, X, and $\underline{t}$ is the same as before (in §2.6), and U denotes a given real function, the "<u>potential</u>", which is defined for all $\underline{t}$ in E^1 and all $(\underline{x}_1, \ldots, \underline{x}_p)$ belonging to some open subset of $(E^n)^p$. Then the system of particles will be called a <u>dynamical system of particles</u>.

If the motions $\underline{t} \to \underline{x}_i(\underline{t})$ of the particles are given, then always a function U can be found such that the equations of motion hold for this particular motion. In general, however, the situation is reverse: the potential function U of the system is given a priori by some extra considerations about the physical nature of the system, and the problem is to determine all the possible motions of the particles or, at least, to characterize them qualitatively from the equation of motion formed with that given function U.

2.9. Thus the problem arises: to solve, for a given potential function U, the system of differential equations

$$\begin{cases} \dfrac{d\dot{X}}{d\underline{t}} = -M^{-1} \cdot U_X(X,\underline{t}) \\ \dfrac{dX}{d\underline{t}} = \dot{X} , \end{cases}$$

or, at least, qualitatively to characterize the solutions. The starting point for any attempt of solving this problem is the fundamental Existence and Uniqueness Theorem of the theory of ordinary differential equations. For formulating it, we need the notion of a connected set: a nonempty subset $\tilde{X}$ of a topological space (X,Γ) is said to be connected if the subspace $(\tilde{X},\tilde{\Gamma})$ of (X,Γ), with the relative topology

$$\tilde{\Gamma} = \{\tilde{X} \cap G | G \in \Gamma\},$$

is a connected space. (Cf. §1.5 and §1.7.)

Let I be a non-empty open interval of real numbers, G be a non-empty connected open subset of the m-dimensional Euclidean number space E^m, g be a mapping from G × I into G, let $\overset{\circ}{t}$ be any point of I, and $\overset{\circ}{y}$ be any point of G. Then, for a sufficiently small open neighbourhood (in I) of $\overset{\circ}{t}$, there exists exactly one solution $t \to y(t)$ of the differential equation

$$\frac{dy}{dt} = g(y,t)$$

which satisfies the "initial condition"

$$y(\overset{\circ}{t}) = \overset{\circ}{y} .$$

This solution, considered as a function of the "starting point" $\overset{\circ}{y}$, the "initial moment" $\overset{\circ}{t}$, and the "time variable" t, has continuous partial derivatives of all orders with respect to all its arguments.

The proof of the first part of the theorem is accomplished by the following steps:

(1) The differential equation together with the initial condition is equivalent to the integral equation

$$y(t) = \overset{\circ}{y} + \int_{\overset{\circ}{t}}^{t} g(y(\tau),\tau)d\tau .$$

(2) Construct, for t sufficiently close to $\overset{\circ}{t}$, a sequence of functions $\{t \rightarrow y^{(n)}(t)\}_{n=0}^{\infty}$ by

$$\begin{cases} y^{(0)}(t) = \overset{\circ}{y}, \\ y^{(n)}(t) = \overset{\circ}{y} + \int_{\overset{\circ}{t}}^{t} g(y^{(n-1)}(\tau),\tau)d\tau, \quad n=1,2,\ldots \end{cases}$$

The definition of the sequence implies that the series

$$\overset{\circ}{y} + \sum_{n=1}^{\infty} (y^{(n)}(t) - y^{(n-1)}(t))$$

converges absolutely and uniformly towards a function $t \rightarrow y(t)$ which satisfies the integral equation and, therefore, both the differential equation and the initial condition.

(3) If there were a second solution, $t \to \tilde{y}(t)$, it would also have to satisfy the integral equation. Hence,

$$\tilde{y}(t) - y(t) = \int_{\overset{\circ}{t}}^{t} \{g(\tilde{y}(\tau),\tau) - g(y(\tau),\tau)\}\, d\tau \ .$$

An application of the Mean-Value Theorem of Differential Calculus to the integrand shows that $\tilde{y} = y$.

For details the reader is referred to the literature.

Note that the existence only of a local solution (t sufficiently close to $\overset{\circ}{t}$) is asserted. If, however, the partial derivatives of the vector function g with respect to the components of the argument vector y are bounded in G, then a local solution $t \to y(t)$ can be extended as long as $y(t)$ does not approach the boundary of G(subject, of course, to the condition that $t \in I$).

2.10. The Existence and Uniqueness Theorem is now applied to dynamical systems of particles:

The derivative $d\underline{x}_i(\underline{t})/d\underline{t}$ is called the velocity vector of the ith particle (relative to the system of coordinates used) at the time $\underline{t}$.

Let $\overset{\circ}{t}$ be a collision-free instant of a dynamical system of particles, $\overset{\circ}{\underline{x}}_1,\ldots,\overset{\circ}{\underline{x}}_p$ be the position vectors and $\overset{\circ}{\dot{\underline{x}}}_1,\ldots,\overset{\circ}{\dot{\underline{x}}}_p$ be the velocity vectors of the particles at the time $\overset{\circ}{t}$. Then the motions of the particles are uniquely determined in some open neighbourhood (in Time) of $\overset{\circ}{t}$.

This neighbourhood is, in general, a subset of the collision-free neighbourhood of $\overset{\circ}{t}$ if to a collision there corresponds a singularity of the potential function U. In exceptional cases, however, a solution of the equations of motion may be extended beyond a moment of collision even if there is a singularity at that moment.

2.11. The order of a system of differential equations can sometimes be reduced by introducing a new system of coordinates such that, in these new coordinates, one of the components of the m-vector solutions becomes constant with respect to the time variable. Thus its time derivative vanishes identically, and the original system of equations is, essentially, reduced to one of only (m-1) equations; thus the problem is simplified. With a view to this technique, transformations of coordinates will be discussed in the following section.

Notes and References

§§2.1 - 2.2. In the history of Mechanics there has been confusion about the concept of mass. The distinction between inertial and gravitational mass, made in the Theory of Relativity, has, in my opinion, only added to it. Newton (1686) defined the mass of a material body extended in Space as the product of volume and mass density. In *this* form, however, the definition is apparently circular, and already Kant (1786) recognized its restricted applicability - namely to homogeneous bodies with identical chemical compositions; consequently, he proposed a new definition based upon the "conservation of momentum" (discussed in §11 of the present text.) This definition, however, has a definite meaning only if the centre of masses of the system of the two bodies compared is taken as being at rest; however the centre of masses can be determined only *after* the masses of the bodies have been determined. Thus Kant's definition also turns out to be circular.

This mistake is avoided in Mach's definition (1883). It is similar to Kant's, but, instead of velocities, accelerations are used for the comparison of masses. (See also Heller 1964). One possible objection (not the only one) is that this definition implicitly pre-supposes the equations of motion and, thus, the notion of mass. In my paper (1962) I gave a definition of "mass" which is based upon Newton's idea, but tries to avoid the circularities mentioned and is logically prior to the equations of motion.

It might be thought preferable to restrict the motions g to collision-free intervals of Time, rather than to have them defined for all instants. Since, however, in Celestial Mechanics a considerable amount of work has been done about collisions (see, e.g., Wintner 1952, Siegel 1956, Siegel-Moser 1971), the more general definition has been given in the text.

It is not necessary to take the notion of particle as primitive. It could, for example, be defined as the ordered pair of a positive real number μ (the mass) and a continuous function $g:T \to S$ (the motion). But such a definition would be artificial and inconvenient: when the function g is restricted to some proper subset T' of T (for instance, to a collision-free Time interval) and $g' \underset{\text{Def.}}{=} g|T'$, then the "particles" (μ,g) and (μ,g') would have to be considered as distinct.

§2.3. For the topological concepts, see the references to §1.

§2.4. As to measurement of time, see Kurth (1964a).

The term "force" has not been used in the text. This, however, does not mean that I think the notion of force to be superfluous in the foundations or applications of Mechanics. But here I take the foundations as already laid. I tried to clarify them in my papers (1964a), (1962), (1960b), (1964b) (listed in the appropriate logical order. Other discussions of the foundations are found, for example, in Mach (1883), Poincaré (1903), Russell (1903), Heller (1964), Suppes (1967), Bunge (1967), Frank (1969).

§2.6. Relatively little seems to be known about the motions of free dynamical systems of p particles, except in the case that p = 2. For Weierstrass' theory of the non-Newtonian two-body problem (i.e., for $\phi(r) \neq r^{-2}$) see Charlier (1902). Results for $p \geq 3$ and the Newtonian law of force are described in Wintner (1952), Siegel (1956), and Siegel-Moser (1971). A few of them have been generalized to unspecified laws of force in Kurth (1955), (1966a), (1966b), (1971a).

§2.9. For ordinary differential equations see, e.g., Lefschetz (1952), Nemytskii-Stepanov (1960), Tricomi (1961), Sansone-Conti (1964), Wilson (1971).

§2.10. For the extension of a solution beyond a moment of collision in systems of gravitating particles, see Wintner (1952), Siegel (1956), Siegel-Moser (1971), Szebehely (1967).

§3. Lagrangian Systems

3.1. There are systems whose particles cannot move freely according to the equations of motion (with a given potential function), but rather are subject to certain constraints given in the form of functional equations.

Consider, for example, a finite set of particles whose motions are constrained by the condition that the mutual distances of any two particles are constant with respect to time. (Thus the particles form a "rigid body".) The equations of the motion, as given in §2.4, do not deal with this case, but rather with the more special case that the number of constraints is zero. Thus the problem arises as to how these equations of motion are to be generalized.

Clearly, this is a physical problem, rather than a mathematical one. But mathematical considerations can assist in devising a reasonable conjecture -- which, then, of course has to be tested empirically. We are here concerned only with the mathematical aspect of the matter.

Thus, for example, one might introduce additional "forces", acting between any two particles, which tend to keep the mutual distances at their initial values; solve the equations of motions; and finally let the magnitudes of these forces in a suitable manner increase indefinitely.

Here another, more formal and more general, method will be persued. New coordinates are introduced such that the restricting equations take a particularly simple form: they express that some of the coordinates are independent of the time variable. It is then assumed that the resulting differential equations give correct descriptions of the motions of the particles.

Thus not only for solving the equations of motion (cf. §2.5), but also for establishing equations which take any given restraints into account, it is useful to study the effects of transformations of coordinates upon the equations of motion.

3.2. The obvious way to do this is to introduce, in the equations of motion of §2.9, unspecified new coordinates. We shall, however, not follow this course here, but shall rather verify the result in a manner which is computationally simpler and has a more general significance: The equations of motion are re-written in a form which will be shown to be invariant under general transformations of local coordinates.

The following reformulation of the equations of motion is immediately verified by the appropriate substitution:

Let

$$\begin{cases} M \frac{d\dot{X}}{d\underline{t}} = - U_X(X,t), \\ \frac{dX}{d\underline{t}} = \dot{X} \end{cases}$$

be the "Newtonian" equations of motion of a dynamical system of particles (cf. §2.9), and define its "Lagrange function" L by

$$L(X,\dot{X},\underline{t}) = \tfrac{1}{2}\, \dot{X}^*M\dot{X} - U(X,\underline{t}).$$

Then the Newtonian equations of motion are equivalent to "Lagrange's equations of motion",

$$\begin{cases} \frac{d}{d\underline{t}} L_{\dot{X}} - L_X = 0 \ , \\ \frac{dX}{d\underline{t}} = \dot{X} \ . \end{cases}$$

"L" stands here as an abbreviation for "$L(X,\dot{X},\underline{t})$, the subscripts "X" and "$\dot{X}$" indicate gradients with respect to X and $\dot{X}$, and "equivalence" means: both sets of equations have exactly the same solutions.

3.3. Now we shall take the second step of the program indicated in the first paragraph of §3.2: to show that equations of the form just introduced are invariant under general transformations of local coordinates. These equations need no longer be the equations of motion of a dynamical system of particles: instead, abstract ordinary differential equations of this particular form will be discussed; the discussion is motivated by the fact that the equations of motion of a dynamical system of particles can be given this form, but the discussion is not restricted to equations of motion.

Let G be a region (i.e. a non-empty open connected subset) of the m-dimensional Euclidean number space E^m, T' be a non-empty open interval of E^1, and $(\underline{x},\dot{\underline{x}},\underline{t}) \rightarrow \underline{L}(\underline{x},\dot{\underline{x}},\underline{t})$ be a real-valued function, called a Lagrange function, which is defined for all $\underline{x} \in \underline{G}$, $\dot{\underline{x}} \in E^m$ and $\underline{t} \in \underline{T}'$. For any fixed value $\underline{t} \in \underline{T}'$, let

$$\underline{\bar{x}} = \underline{\bar{\phi}}(\underline{x},\underline{t})$$

be a transformation of local coordinates such that $\bigcap \{\underline{\bar{\phi}}(\underline{G},\underline{t}) \,|\, \underline{t} \in \underline{T}'\}$ contains a region $\underline{\bar{G}}$ of E^m, and let

$$\underline{x} = \underline{\phi}(\underline{\bar{x}},\underline{t})$$

be the local inverse of $\underline{\bar{\phi}}$ on $\underline{\bar{G}}$. Thus, for any fixed value $\underline{t}$ of $\underline{T}'$, the mappings $\underline{\bar{\phi}}:\underline{G} \rightarrow \underline{\bar{G}}$ and $\underline{\phi}:\underline{\bar{G}} \rightarrow \underline{G}$ are defined on domains which are independent of the parameter $\underline{t}$.

We now define the transform $\underline{\bar{L}}$ of the Lagrange function $\underline{L}$ under the coordinate transformation $\underline{\bar{\phi}}$ by

$$\underline{\bar{L}}(\underline{\bar{x}},\dot{\underline{\bar{x}}},\underline{t}) = \underline{L}[\underline{\phi}(\underline{\bar{x}},\underline{t}),\ \underline{\phi}_{\underline{\bar{x}}^*}(\underline{\bar{x}},\underline{t})\dot{\underline{\bar{x}}}+\underline{\phi}_{\underline{t}}(\underline{\bar{x}},\underline{t}),\underline{t}] \quad \text{for any } \underline{\bar{x}} \in \underline{\bar{G}},\ \dot{\underline{\bar{x}}} \in E^m,\ \underline{t} \in \underline{T}'$$

(or, in short, by

$$\underline{\bar{L}} = \underline{L}(\underline{\phi},\underline{\phi}_{\underline{\bar{x}}^*}\dot{\underline{\bar{x}}} + \underline{\phi}_{\underline{t}},\underline{t})$$

where $\underline{\phi}_{\underline{\bar{x}}^*}$ stands for the functional matrix $\{\partial\phi^i/\partial\bar{x}^j\}_{i,j=1}^{m}$ of $\underline{\phi} = \{\phi^i\}_{i=1}^{m}$, and $\underline{\phi}_{\underline{t}}$ for $\{\partial\phi^i/\partial t\}_{i=1}^{m}$. Similar abbreviations will be used in the following without formal introductions. Evidently every Lagrange function is the transform of itself under the identity transformation.

3.4. If $\underline{\bar{L}}$ is the transform of the Lagrange function $\underline{L}$ under the coordinate transformation $\underline{\bar{\phi}}$, then $\underline{L}$ is the transform of the Lagrange function $\underline{\bar{L}}$ under the inverse coordinate transformation $\underline{\phi}$.

The assertion is symbolized by the diagram

$$L \;\; \substack{\phi \\ \leftarrow \\ \rightarrow \\ \bar{\phi}} \;\; \bar{L} \; .$$

It is proven by a straight-forward application of the transform ϕ to the Lagrange function $\bar{L}$. The proof of the following proposition is similar:

3.5. <u>If $\bar{L}$ is the transform of the Lagrange function L under the coordinate transformation $\bar{\phi}$, and $\bar{\bar{L}}$ is the transform of the Lagrange function $\bar{L}$ under the coordinate transformation $\bar{\bar{\phi}}$, then $\bar{\bar{L}}$ is the transform of the Lagrange function L under the composite coordinate transformation $\bar{\bar{\phi}}\,\bar{\phi}$</u> :

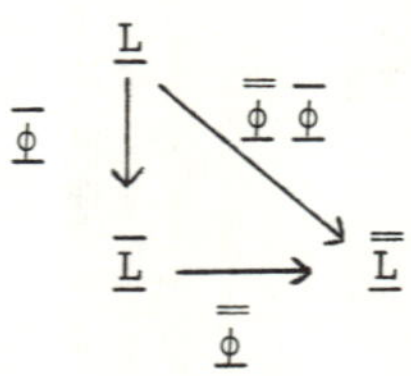

Therefore, by this and the preceding results, <u>the relation "is the transform of the Lagrange function ··· under some coordinate transformation" is an equivalence relation.</u>

3.6. We now turn to the general form of the left-hand sides of Lagrange equations as introduced in §3.2.

Let $f : T' \to G$ be any <u>curve</u>, or <u>motion</u>, in G, i.e., any function from T' into G. Then the function $\Lambda_L f : T' \to E^m$, defined by

$$(\Lambda_L f)(t) \underset{\text{Def.}}{=} \frac{d}{dt}\left(L_{\dot{x}}\left(f(t), df(t)/dt, t\right)\right) - L_x\left(f(t), df(t)/dt, t\right)$$

or, in short,

$$\Lambda_{\underline{L}}\,\underline{f} \underset{\text{Def.}}{=} \frac{d}{d\underline{t}}\left(\underline{L}_{\dot{\underline{x}}}\left(\underline{f}, d\underline{f}/d\underline{t}, \underline{t}\right)\right) - \underline{L}_{\underline{x}}\left(\underline{f}, d\underline{f}/d\underline{t}, \underline{t}\right),$$

is called the Lagrange derivative of the function $\underline{f}$ with respect to the Lagrange function $\underline{L}$.

For a dynamical system of particles, the Lagrange derivative of any motion of the particles with respect to the Lagrange function of the system vanishes identically, by §3.2, if the system of Cartesian coordinates of §3.2 is used. It will now be shown that this is true for any admissible system of local coordinates. The assertion is an immediate consequence of the following proposition:

3.7. Assume the hypotheses and notation of §3.3 - §3.6, and let

$$\overline{\underline{f}}\,(\underline{t}) \underset{\text{Def.}}{=} \overline{\underline{\phi}}\left(\underline{f}(\underline{t}), \underline{t}\right) \qquad \text{for } \underline{t} \in \underline{T}' .$$

Then,

$$\left(\Lambda_{\overline{\underline{L}}}\,\overline{\underline{f}}\right)(\underline{t}) = \underline{\phi}_{\overline{\underline{x}}^*}\,(\overline{\underline{x}}, \underline{t}) \cdot \left(\Lambda_{\underline{L}}\,\underline{f}\right)(\underline{t}) \ , \ \underline{t} \in \underline{T}'$$

or, for short,

$$\Lambda_{\overline{\underline{L}}}\,\overline{\underline{f}} = \underline{\phi}_{\overline{\underline{x}}^*} \cdot \Lambda_{\underline{L}}\,\underline{f} \, .$$

For coordinate transformations $\overline{\underline{\phi}}$ which are independent of the time variable $\underline{t}$ we shall also use the notation

$$\left(\Lambda_{\overline{\underline{L}}}\,\overline{\underline{f}}\right)_i = \frac{\partial x^j}{\partial \overline{x}^i}\left(\Lambda_{\underline{L}}\,\underline{f}\right)_j$$

where $\left(\Lambda_{\overline{\underline{L}}}\,\overline{\underline{f}}\right)_i$ denotes the ith component of $\Lambda_{\overline{\underline{L}}}\,\overline{\underline{f}}$, and the summation convention has been applied by which equal sub- and superscripts within a term indicate summation over this index from 1 to m. Thus, for example, $\alpha_k\beta^k$ stands for $\sum_{k=1}^{m} \alpha_k\beta^k$.

The proposition is proven by straight-forward applications of the definitions of $\overline{\underline{L}}$, $\overline{\underline{f}}$, and $\Lambda_{\overline{\underline{L}}}\,\overline{\underline{f}}$ for given $\underline{L}$, $\underline{f}$, and $\overline{\underline{\phi}}$.

Thus, <u>Lagrange equations are invariant under any transformation of local coordinates</u>.

<u>3.8.</u> Now it is possible to give the terms "Lagrange function", "Lagrange derivative", and "Lagrange curve" meanings which are independent of the coordinate system chosen. By "<u>Lagrange curve</u>" we mean a solution $\underline{t} \to \underline{x}(\underline{t})$ of a given Lagrange equation. To show this independence, we make the following definitions:

Let M be an m-dimensional analytical manifold, G be a coordinate neighbourhood in M, $\underline{h}:G \to E^m$ be a system of local coordinates in G, $\underline{T}'$ be an open interval in E^1 which contains the point (instant) $\underline{t} = 0$, $f:\underline{T}' \to G$ be a curve in G, and $\mathring{x} = f(0)$. To the curve f in G there corresponds the curve $\underline{h}f$ in E^m, and this curve has, at the point $\underline{\mathring{x}} \underset{\text{Def.}}{=} \underline{h}(\mathring{x})$, the tangent vector

$$\underline{\lambda}(\underline{\mathring{x}}) \underset{\text{Def.}}{=} \left(\frac{d}{dt}\,\underline{h}f \right)(t)\Big|_{\underline{t}=0} .$$

Now introduce in G a second system of coordinates, $\overline{\underline{h}}:G \to E^m$, and let $\overline{\underline{\phi}} \underset{\text{Def.}}{=} \underline{\overline{h}h^{-1}}$ be the coordinate transformation $\underline{x} \to \overline{\underline{x}}$ from $\underline{h}(G)$ onto $\overline{\underline{h}}(G)$. The curve $\overline{\underline{h}}f$ has, at the point $\overline{\underline{\mathring{x}}} = \overline{\underline{h}}(\mathring{x})$, the tangent vector

$$\overline{\underline{\lambda}}(\underline{x}) = \left(\frac{d}{dt}\,\overline{\underline{h}}f \right)(t)\Big|_{\underline{t}=0} ,$$

and since

$$\overline{\underline{h}}f = (\overline{\underline{\phi}}\,\underline{h})f = \overline{\underline{\phi}}(\underline{h}f),$$

both tangent vectors are related by

$$\overline{\underline{\lambda}}(\overline{\underline{\mathring{x}}}) = \overline{\underline{\phi}}_{\underline{x}*}(\underline{\mathring{x}})\;\underline{\lambda}(\underline{\mathring{x}})$$

or, in a short-hand notation,

$$\underline{\overline{\lambda}}^{i} = \frac{\partial \overline{x}^{i}}{\partial x^{j}} \lambda^{j} \qquad \text{at } \underline{x} = \underline{\overset{o}{x}},\ \underline{\overline{x}} = \underline{\overline{\overset{o}{x}}}\text{, resp.}$$

(As to the notation, see §3.3, and §3.7.)

3.9. Now let A be any given non-empty subset of the manifold M, and G be a coordinate neighbourhood on M which intersects A, and suppose that to each admissible system of local coordinates $\underline{h}$ in G there corresponds a vector function $\underline{\lambda}:\underline{h}(G \cap A) \to E^m$ such that, for any two admissible systems of coordinates $\underline{h}$, $\underline{\overline{h}}$ in G and the corresponding vector functions $\underline{\lambda}$, $\underline{\overline{\lambda}}$,

$$\underline{\overline{\lambda}}(\underline{\overline{x}}) = \underline{\overline{\phi}}_{\underline{x}*}(\underline{x})\ \underline{\lambda}(\underline{x})$$

or, for short,

$$\overline{\lambda}^{i} = \frac{\partial \overline{x}^{i}}{\partial x^{j}} \lambda^{j}$$

for any $x \in G \cap A$, and any neighbourhood G intersecting A. Then we shall (temporarily) say the function $\underline{\overline{\lambda}}$ is <u>equivalent to</u> the function $\underline{\lambda}$. This relation "is equivalent to" is an equivalence relation, and thus it defines, for each point x of A, an equivalence class, $\lambda(x)$ (say). We call the function $x \to \lambda(x)$ a <u>contravariant vector field</u> defined on the set A, and its value $\lambda(x)$ at the arbitrary point x of A a <u>contravariant vector</u> at the point x. Any member $\underline{\lambda}(\underline{x})$ of the equivalence class $\lambda(x)$ is said to be a <u>representative</u> of the contravariant vector $\lambda(x)$, and any function $\underline{x} \to \underline{\lambda}(\underline{x})$ defined by covering the point x by an admissible system of local coordinates will be called a <u>local representative</u> of the contravariant vector field $x \to \lambda(x)$.

3.10. Suppose that, for two given contravariant vector fields defined in the same domain, there is a covering of the domain by a system of local coordinate systems with respect to which the local representatives of both vector fields are equal. Then the local representatives of the fields are equal with respect to every admissible system of local coordinates. In such a case the vector fields themselves are <u>equal</u>, by their definition as equivalence classes.

If, in particular, some representative of a contravariant vector, or vector field, vanishes, then each of its representatives vanishes. We shall then say: the vector (field) vanishes; or: it is the <u>null vector (field)</u>, which will be denoted by 0.

Let $x \to \lambda(x)$ and $x \to \mu(x)$ be two contravariant vector fields defined in the same domain, $\underline{x} \to \underline{\lambda}(\underline{x})$ and $\underline{x} \to \underline{\mu}(\underline{x})$ be their representatives with respect to some admissible system of local coordinates, and α and β be any two real numbers. Then the equivalence class to which the function $\underline{x} \to \alpha\underline{\lambda}(\underline{x}) + \beta\underline{\mu}(\underline{x})$ belongs is independent of the choice of the system of local coordinates, and depends only on the vector fields λ and μ, and the numbers α and β. It will be denoted by $\alpha\lambda + \beta\mu$. The contravariant vector field $\alpha\lambda$ is called a <u>scalar multiple</u> of the contravariant vector field λ, and the vector field $\lambda + \mu$ is said to be the <u>sum</u> of the vector fields and μ.

The algebraic rules governing scalar multiples and sums are obviously those of a real vèctor space.

<u>3.11</u>. In §3.6–§3.9 it was shown that the tangent vector $\underline{\lambda}(\underline{x})$ constructed there is the representative of a contravariant vector. We shall call the contravariant vector $\lambda(x)$ which is determined by a tangent vector $\underline{\lambda}(\underline{x})$ (for a given curve passing through the given point $\overset{\circ}{x}$ of the manifold) a <u>contravariant tangent vector</u> of the manifold at the point $\overset{\circ}{x}$ or briefly, if no misunderstanding appears likely, a <u>tangent vector</u>.

The set of all tangent vectors assigned to the point x of the manifold (i.e., the set of the tangent vectors at x of all the curves passing through x) will be called the <u>tangent (vector) space</u> of M at x, and denoted by $\dot{M}(x)$. Clearly, if M is m-dimensional, then $\dot{M}(x)$ is, at each point x of M, also m-dimensional. That is, all tangent spaces are isomorphic to the m-dimensional Euclidean number space E^m.

<u>3.12</u>. Now we are in the position to define a Lagrange function L independently of any particular system of coordinates. Let M be a connected m-dimensional analytic manifold, $\dot{M}(x)$ be its tangent space at any point x of M, T' be a nonempty open time interval, and L be a given real function of $(x, \dot{x}, t)$ where x is any point of M, $\dot{x}$ is any point of $\dot{M}(x)$, and t is any moment in T'. Then L will be called a <u>Lagrange function</u> on the manifold M, and M, together with L, will be said to be a <u>Lagrange manifold</u>. The local representative of L with respect to a given coordinate system $x \to \underline{x}$ is defined by

$$\underline{L}(\underline{x},\dot{\underline{x}},\underline{t}) = L(x,\dot{x},t)$$

where $\underline{x}$, $\dot{\underline{x}}$, $\underline{t}$ are the coordinate vectors of x, $\dot{x}$, t, respectively. For any other system of local coordinates $x \to \overline{\underline{x}}$ the local representative of the Lagrange function L is given by

$$\overline{\underline{L}}(\overline{\underline{x}},\dot{\overline{\underline{x}}},\underline{t}) = L(x,\dot{x},t),$$

and it is easily verified that <u>$\overline{\underline{L}}$ is the transform of $\underline{L}$ under the coordinate transformation $\underline{x} \to \overline{\underline{x}}$</u>. In the form of a diagram:

$$\begin{array}{ccc} & L & \xrightarrow{\;x \to \overline{\underline{x}}\;} \searrow \\ x \to \underline{x} & \downarrow & \\ & \underline{L} & \xrightarrow[\underline{x} \to \overline{\underline{x}}]{} \overline{\underline{L}} \end{array}$$

This also holds for time-dependent coordinate transformations.

3.13. Having shown that the notion of a Lagrange function has a meaning which is independent of the particular system of local coordinates used, we now wish to show that the same also holds for the Lagrange derivative of a curve. For this purpose we introduced the concepts of a covariant vector and a covariant vector field, in a way similar to that of §3.9:

Again, let A be a given non-empty subset of the m-dimensional analytical manifold M, and G be a coordinate neighbourhood on M which intersects A. Suppose that to each admissible system of local coordinates $\underline{h}$ there corresponds a vector function $\underline{\mu}:\underline{h}(G \cap A) \to E^m$ such that, for any two admissible systems of coordinates $\underline{h}$, $\overline{\underline{h}}$ in G and the corresponding vector functions $\underline{\mu}$, $\overline{\underline{\mu}}$,

$$\overline{\underline{\mu}}(\overline{\underline{x}}) = \underline{\phi}_{\underline{x}^*}{}^*(\overline{\underline{x}})\underline{\mu}(\underline{x})$$

where

$$\underline{\phi} \underset{\text{Def.}}{=} \underline{h}\overline{\underline{h}}^{-1}$$

or, in short,

$$\overline{\mu}_i = \frac{\partial x^j}{\partial \overline{x}^i} \mu_j$$

for any x ε G ∩ A, and any neighbourhood G intersecting A. Then we shall (temporarily) say the function $\overline{\underline{\mu}}$ is equivalent to the function $\underline{\mu}$. The relation "is equivalent to" is an equivalence relation, and thus it defines, for each point x of A, an equivalence class, μ(x) (say). We call the function x → μ(x) a covariant vector field defined on the set A, and its value μ(x) at the arbitrary point x of A- a covariant vector at the point x. Any member $\underline{\mu}(\underline{x})$ of the equivalence class μ(x) is called a representative of the covariant vector μ(x), and any function $\underline{x} \to \underline{\mu}(\underline{x})$ defined by covering the point x by an admissible system of local coordinates will be called a <u>local representative</u> of the covariant vector field $x \to \mu(x)$.

The algebraic rules of covariant vector fields are the same as those of contravariant vector fields. (Cf. §3.10).

3.14. Let $f:\underline{T}' \to M$ be a curve on the Lagrange manifold M, and suppose that the mapping f is one-to-one. Further let $x \to \underline{x}$ and $x \to \bar{\underline{x}}$ be two admissible systems of coordinates defined in a coordinate neighbourhood in M which intersects the range of f, and $\underline{f}(\underline{t})$ and $\bar{\underline{f}}(\underline{t})$ be the coordinate vectors of points $f(\underline{t})$ with respect to these systems of local coordinates. Then, by §3.7,

$$\left(\Lambda_{\bar{\underline{L}}}\, \bar{\underline{f}}\right)_i = \frac{\partial x^j}{\partial \bar{x}^i} \left(\Lambda_{\underline{L}}\, \underline{f}\right)_j$$

Thus <u>the Lagrange derivative of $\underline{f}$ and $\bar{\underline{f}}$ can be considered as local representatives of a covariant vector field, defined on the range</u> $\{f(\underline{t}) \mid \underline{t} \in \underline{T}'\}$ <u>of the motion</u> f. We shall call this covariant vector field the <u>Lagrange derivative</u> of the motion f with respect to the Lagrange function L, and denote it by $\Lambda_L f$. It has a meaning independent of the system of coordinates used.

The independence is graphically described by the diagram

$$\begin{array}{ccc} \underline{L} & \xrightarrow[\bar{\underline{\phi}}]{} & \bar{\underline{L}} \\ \Lambda_{\underline{L}} \downarrow & & \downarrow \Lambda_{\bar{\underline{L}}} \\ \Lambda_{\underline{L}}\, \underline{f} & \xrightarrow[\bar{\underline{\phi}}]{} & \Lambda_{\bar{\underline{L}}}\, \bar{\underline{f}} \end{array}$$

A Lagrange curve $\underline{f}$ has, in §3.8, been defined as a curve whose Lagrange derivative vanishes. Since this condition also has a coordinate-independent meaning, the notion of a Lagrange curve on the manifold M can be defined: it is a curve on M whose Lagrange derivative vanishes identically.

3.15. In all these coordinate transformations we have transformed only the local coordinates on the manifold M and, in the manner prescribed by these transformations, the coordinates on the tangent spaces $\dot{M}(x)$. But we have not transformed the coordinates of Time. In fact, a short straight-forward computation shows that a Lagrange derivative is transformed by a transformation of the time coordinate into a Lagrange derivative if, and only if, the transformation is linear--that is if (and only if) the time scale and the time zero are changed. Thus, on Time, there is a set of distinguished coordinate systems, namely those obtained by linear transformation from a system with respect to which the Lagrange functions and derivatives have been formed. Since these distinct systems of time coordinates are not essentially different, we choose any one of them once for all as the system of coordinates. In consequence, from now on we shall identify Time with the 1-dimensional Euclidean number-space, E^1, and the instants t with their coordinates t, supposing that the coordinates have been chosen in such a way that Lagrange's equation of motion hold for mechanical systems of particles.

3.16. We now return to the problem posed in §3.1: to set up, hypothetically, equations of motion for a system of particles of which the possib positions (and, therefore, motions) are subject to certain given constraints. We proceed by the following steps:

(i) First suppose there are no constraints. Then, by §3.2, there is for the unrestrained system a Lagrange function which is defined in some region G of a Euclidean number space $(E^n)^p$ (the configuration space) and the corresponding tangent spaces.

(ii) Introduce the constraints and suppose that their effect is to restrict the configuration vector x of the system to some m-dimensional manifold M situated in G.

(iii) The Lagrange function considered in (i) "induces" a Lagrange function L on M: it is the former Lagrange function, restricted to M and the tangent spaces $\dot{M}(x)$, $x \in M$, (to the "tangent bundle" $\{(x,\dot{x}) | x \in M, \dot{x} \in \dot{M}(x)\}$).

(iv) Generalizing the result of §3.6, we postulate, tentatively, that for any motion f of the system subject to the constraints the Lagrange derivative $\wedge_L f$ vanishes identically.

By §3.14, this equation of motion, $\wedge_L f = 0$, has a meaning independent of the particular choice of the coordinates by the constancy of which the constraints are described.

3.17. The discussion in §3.16 referred to systems of particles. It can be generalized in the following way:

Take the terms "dynamical system" and "configuration" (of a given dynamical system) as primitive. To any dynamical system a set of possible configurations is assigned. If this set is a Lagrange manifold M and any "motion" of the system $f: T' \to M$ (where T' is some non-empty open Time interval) is a Lagrange curve on M, then the system is called a Lagrange system. If m is the dimension of M, the system is said to have m degrees of freedom.

3.18. For any given system of local coordinates the Lagrange equations of a given Lagrange system read

$$\frac{d}{dt} L_{\underline{\dot{x}}} - L_{\underline{x}} = 0, \quad \frac{d}{dt} \underline{x} = \underline{\dot{x}} ,$$

or, equivalently,

$$\underline{L}_{\dot{\underline{x}}\dot{\underline{x}}^*}\frac{d\dot{\underline{x}}}{dt} + \underline{L}_{\dot{\underline{x}}\underline{x}^*}\frac{d\underline{x}}{dt} + \underline{L}_{t\dot{\underline{x}}} - \underline{L}_{\underline{x}} = 0, \quad \frac{d}{dt}\underline{x} = \dot{\underline{x}}$$

where again subscripts indicate differentiations. Symbols like $\underline{L}_{\dot{\underline{x}}\dot{\underline{x}}^*}$ and $\underline{L}_{\dot{\underline{x}}\underline{x}^*}$ stand for matrices of second order derivatives - in the present cases for $\{\partial^2\underline{L}/\partial\dot{x}^i\partial\dot{x}^j\}_{i,j=1}^{n}$ and $\{\partial^2\underline{L}/\partial\dot{x}^i\partial x^j\}_{i,j=1}^{n}$, respectively.

The local Lagrange equations can be given the standard form of ordinary differential equations

$$dy/dt = g(y,t)$$

(cf. §2.9) if and only if the matrix $\underline{L}_{\dot{\underline{x}}\dot{\underline{x}}^*}$ is non-singular.

3.19. A straight-forward calculation, using the definition of the transform $\overline{\underline{L}}$ of the local Lagrange function $\underline{L}$ (cf. §3.3), shows that

$$\frac{\partial^2\overline{\underline{L}}}{\partial\dot{\overline{x}}^i\,\partial\dot{\overline{x}}^j} = \frac{\partial\dot{x}^k}{\partial\dot{\overline{x}}^i}\;\frac{\partial\dot{x}^\ell}{\partial\dot{\overline{x}}^j}\;\frac{\partial^2\underline{L}}{\partial\dot{x}^k\,\partial\dot{x}^\ell}$$

for any (time-independent) admissible transformation of local coordinates $\underline{x} \rightarrow \overline{\underline{x}}$. This implies that

$$\frac{\partial^2\overline{\underline{L}}}{\partial\dot{\overline{x}}^i\,\partial\dot{\overline{x}}^j} = \frac{\partial x^k}{\partial\overline{x}^i}\;\frac{\partial x^\ell}{\partial\overline{x}^j}\;\frac{\partial^2\underline{L}}{\partial\dot{x}^k\,\partial\dot{x}^\ell}$$

since

$$\dot{x}^k = \frac{\partial x^k}{\partial\overline{x}^i}\,\dot{\overline{x}}^i$$

and, therefore,

$$\frac{\partial\dot{x}^k}{\partial\dot{\overline{x}}^i} = \frac{\partial x^k}{\partial\overline{x}^i}$$

The same transformation of $L_{\dot{x}\dot{x}^*}$ holds also for time-dependent coordinate transformations.

Hence, if at a point $(x, \dot{x}, t)$ the Hessian matrix $L_{\dot{x}\dot{x}^*}$ of some local representative L of the Lagrange function L is non-singular, then the Hessian matrix $\bar{L}_{\dot{\bar{x}}\dot{\bar{x}}^*}$ of any local representative $\bar{L}$ of L is non-singular at $(x,\dot{x},t)$. Note that even the sign of det $L_{\dot{\bar{x}}\dot{\bar{x}}^*}$ is invariant under coordinate transformations $x \to \bar{x}$.

Thus the validity of the inequality det $L_{\dot{x}\dot{x}^*} \neq 0$ is a property of the Lagrange function L, rather than of its representation L. A Lagrange system whose Lagrange function has this property will be called non-degenerate. All Lagrange systems discussed in the following will tacitly be assumed to be non-degenerate.

3.20. Finally it will be shown that the Hessian matrix $L_{\dot{\bar{x}}\dot{\bar{x}}^*}$ can be given a coordinate-independent meaning. For this end we introduce the notions of covariant and contravariant tensor fields of the second order, in analogy to those of covariant and contravariant vector fields as given in §3.13 and §3.8.

Let A be a given non-empty subset of an m-dimensional analytic manifold M which intersects a coordinate neighbourhood G in M, and suppose that to each admissible system of local coordinates h in G there corresponds a (m×m)-matrix function $\mu : h(G \cap A) \to E^{(m^2)}$ such that, for any two admissible system of coordinates $h, \bar{h}$ in G and the corresponding matrix functions μ, $\bar{\mu}$,

$$\bar{\mu}(\bar{x}) = \phi_{\bar{x}^*}{}^*(\bar{x})\mu(x)\phi_{\bar{x}^*}(\bar{x})$$

where

$$\underline{\phi} \underset{\text{Def.}}{=} \underline{h}\,\underline{\bar{h}}^{-1}$$

or, in short,

$$\bar{\mu}_{ij} = \frac{\partial x^k}{\partial \bar{x}^i}\frac{\partial x^\ell}{\partial \bar{x}^j}\mu_{k\ell}$$

for any $x \in G \cap A$, and any neighbourhood G intersecting A. Then we shall (temporarily) say the function $\underline{\bar{\mu}}$ _is equivalent to_ the function $\underline{\mu}$. This relation "is equivalent to" is an equivalence relation, and thus it defines, for each point x of A, an equivalence class, $\mu(x)$ (say). The function $x \to \mu(x)$ is called a _covariant tensor field of the second order_ on A, and its value $\mu(t)$ at the arbitrary point x of A is said to be _covariant tensor of the second order_. Any member $\underline{\mu}$ of the equivalence class μ is said to be a representative of this class, and similarly the class $\mu(x)$ is _represented_ by any of its members, $\underline{\mu}(x)$ (say).

3.21. In a similar way, contravariant and mixed tensors and tensor fields of the second and higher orders are defined. Thus, for example, a _contravariant tensor field_ of the second order is defined by means of the equivalence relation

$$\bar{\lambda}^{ij} = \frac{\partial \bar{x}^i}{\partial x^k}\frac{\partial \bar{x}^j}{\partial x^\ell}\lambda^{k\ell}$$

between its representatives, and the corresponding transformation rule for the representatives of a _mixed tensor field_ ν of the second order reads

$$\bar{\nu}_i^{\,j} = \frac{\partial x^k}{\partial \bar{x}^i}\frac{\partial \bar{x}^j}{\partial x^\ell}\nu_k^{\,\ell} .$$

The algebraic rules for tensor fields are those of a (real) vector space. For details, see §3.10.

<u>3.22</u>. Now let $t \to f(t)$ be a curve on the Lagrange manifold M, and suppose that the mapping f is one-to-one. Restrict the Hessian $\underline{L}_{\dot{\underline{x}}\,\dot{\underline{x}}^*}$, of any local representative $\underline{L}$ of the Lagrange function L to the range of the motion f, or rather its local representative $\underline{f}$. Then, by §3.18 - §3.20, <u>the Hessian</u> $\underline{L}_{\dot{\underline{x}}\,\dot{\underline{x}}^*}$ <u>can be interpreted as a local representative of a co-variant tensor field</u> $L_{\dot{x}\,\dot{x}^*}$ <u>of the second order, defined on the range of the motion</u> f.

Notes and References

<u>§3.1</u>. For other methods of generalizing Newton's equations of motion, see, for example, Whittaker (1961) or Joos (1959).

<u>§3.2</u>. The derivation of Lagrange's equation from Newton's is found, for example, in Whittaker (1961).

<u>§§3.12-13</u>. The notion of a tensor field as introduced in the text is close to the notion used in Physics. For another definition (which is the more usual one in mathematics) see, for example, Laugwitz (1965). For basic concepts of differential topology see Portrjagin (1958), Munkres (1966).

<u>§3.19</u>. Degenerate Lagrange systems have been discussed by Shanmugadhasan (1963).

CHAPTER II

HAMILTON-JACOBI THEORY

§4. Hamilton's Equations

4.1. By §3.18 the local Lagrange equations

$$\frac{d}{dt} L_{\dot{x}} - L_x = 0, \quad \frac{dx}{dt} = \dot{x}$$

can be given the standard form of ordinary differential equations. There is a second way to obtain this form: Define m-vectors y by

$$y = L_{\dot{x}} \; (x, \dot{x}, t) \; ;$$

thus the first Lagrange equation is transformed into the differential equation

$$\frac{dy}{dt} = L_x \; (x, \dot{x}, t)$$

where now $\dot{x}$ is to be considered as a function of the independent variables x, y, and t. This function is (locally) uniquely determined by the above equation defining y , on account of the:

Inverse-Mapping Theorem. Let $f : U \to E^m$ be a mapping from the open subset U of the m-dimensional Euclidean number space E^m into E^m, and $\overset{o}{x}$ be a point of U at which the Jacobian matrix f_{x^*} of f does not vanish. Then there is an open neighbourhood $\overset{o}{U} \subset U$ of $\overset{o}{x}$ such that $f|\overset{o}{U}$, the restriction of f to $\overset{o}{U}$, has an inverse $(f \,|\, \overset{o}{U})^{-1}$ defined in an open neighbourhood of $f(\overset{o}{x})$. This "local inverse of f" is unique, and its Jacobian matrix $(f^{-1})_y$ satisfies the equation

$$(f^{-1})_{y^*} \cdot f_{x^*} = I_m$$

where $\underline{y} = \underline{f}(\underline{x})$.

I_m denotes the (m×m) unit matrix.

4.2. The equation of §4.1 for $d\underline{y}/dt$ can be given explicitly:

Let $\underline{L}$ be a local Lagrange function, and define a function $\underline{H}$ of $(\underline{x},\underline{y},t)$, the local "Hamilton function" corresponding to the local Lagrange function $\underline{L}$, in the appropriate domain by

$$\underline{H}(\underline{x},\underline{y},t) + \underline{L}(\underline{x},\dot{\underline{x}},t) = \underline{y}^*\dot{\underline{x}}$$

where

$$\underline{y} = \underline{L}_{\dot{\underline{x}}}(\underline{x},\dot{\underline{x}},t). \qquad \text{(Def.)}$$

(Thus, as in §4.1, $\dot{\underline{x}}$ is to be considered as a function of $(\underline{x},\underline{y},t)$. Then the local Lagrange equations

$$\begin{cases} \dfrac{d}{dt}\underline{L}_{\dot{\underline{x}}}(\underline{x},\dot{\underline{x}},t) - \underline{L}_{\underline{x}} = 0 \,, \\[2mm] \dfrac{d\underline{x}}{dt} = \dot{\underline{x}} \,, \end{cases}$$

and the "local Hamilton equations"

$$\begin{cases} \dfrac{d\underline{x}}{dt} = \underline{H}_{\underline{y}}(\underline{x},\underline{y},t) \,, \\[2mm] \dfrac{d\underline{y}}{dt} = -\underline{H}_{\underline{x}}(\underline{x},\underline{y},t) \end{cases}$$

are equivalent; that is: any solution of one system is transformed into a solution of the other one by the transformation

$$\begin{cases} t = t \,, \\ \underline{x} = \underline{x} \,, \\ \underline{y} = \underline{L}_{\dot{\underline{x}}}(\underline{x},\dot{\underline{x}},t) \end{cases}$$

<u>or its inverse</u>

$$\begin{cases} t = t \quad , \\ \underline{x} = \underline{x} \quad , \\ \dot{\underline{x}} = \underline{H}_{\underline{y}}(\underline{x},\underline{y},t), \end{cases}$$

<u>resp</u>.

The proof of this proposition will be based upon the following lemma.

4.3. Let A be a real function of the m-vector variables ξ defined in an open connected neighbourhood of some point $\overset{o}{\xi}$ of E^m ; suppose that

$$\det A_{\xi\xi*}(\overset{o}{\xi}) \neq 0 ,$$

and let

$$\eta \underset{\text{Def.}}{=} A_\xi(\xi) .$$

Then the mapping A_ξ has an unique inverse in some open connected neighbourhood of $\overset{o}{\eta} \underset{\text{Def.}}{=} A_\xi(\overset{o}{\xi})$, given by

$$\xi = B_\eta(\eta)$$

where B, as a function of η, is defined by

$$A(\xi) + B(\eta) = \xi^* \eta$$

in an appropriate neighbourhood of $\overset{o}{\eta}$.

The mappings $\eta \to \xi$ and $\xi \to \eta$ are called <u>Legendre transformations</u>.

<u>Proof</u>. The existence of the mapping follows from the Inverse-Mapping Theorem of §4.1. To prove its particular form, differentiate the equation defining B with respect to η; thus,

$$\frac{\partial A}{\partial \xi^j}\frac{\partial \xi^j}{\partial \eta_i} + \frac{\partial B}{\partial \eta_i} = \frac{\partial \xi^j}{\partial \eta_i}\eta_j + \xi^i .$$

This, together with the definition of η, yields the assertion.

<u>Corollary</u> $$A_{\xi\xi*} \cdot B_{\eta\eta*} = I_m .$$

4.4. Proof of Proposition 4.2:

(1) The definitions

$$\begin{cases} \underline{H}(\underline{x},\underline{y},t) + \underline{L}(\underline{x},\dot{\underline{x}},t) = \underline{y}^* \dot{\underline{x}} , \\ \underline{y} = \underline{L}_{\dot{\underline{x}}}(\underline{x},\dot{\underline{x}},t) \end{cases}$$

and the hypothesis of §3.19 that det $\underline{L}_{\dot{\underline{x}}\dot{\underline{x}}^*} \neq 0$ imply, by Lemma 4.3, that

$$\dot{\underline{x}} = \underline{H}_{\underline{y}}(\underline{x},\underline{y},t)$$

(2) Hence the first Hamilton equation,

$$\frac{dx}{dt} = \underline{H}_{\underline{y}}(\underline{x},\underline{y},t)$$

(3) The Lagrange equation

$$\frac{d}{dt} \underline{L}_{\dot{\underline{x}}} - \underline{L}_{\underline{x}} = 0$$

and the definition of $\underline{y}$ imply that

$$\frac{d\underline{y}}{dt} = \underline{L}_{\underline{x}}(\underline{x},\dot{\underline{x}},t).$$

(4) $\underline{L}_{\underline{x}}(\underline{x},\dot{\underline{x}},t) = (\underline{y}^* \dot{\underline{x}} - \underline{H}(\underline{x},\underline{y},\underline{t}))_{\underline{x}}$

where $\underline{y}$ is to be considered as a function of $\underline{x}$, $\dot{\underline{x}}$, and t.

(5) Hence,

$$\underline{L}_{\underline{x}}(\underline{x},\dot{\underline{x}},t) = \underline{y}_{\underline{x}}{}^* \dot{\underline{x}} - \underline{H}_{\underline{x}} - \underline{y}_{\underline{x}}{}^* \underline{H}_{\underline{y}} = -\underline{H}_{\underline{x}} , \qquad (1)$$

and

$$(6)\quad \begin{cases} \dfrac{d\underline{x}}{dt} = \underline{H}_{\underline{y}}\ , & (2) \\[2ex] \dfrac{d\underline{y}}{dt} = -\underline{H}_{\underline{x}}\ . & (3)\ (5) \end{cases}$$

Thus it is shown that the solutions of the Lagrange equations are transformed into solutions of the corresponding Hamilton equations. The converse is proven in a similar way.

4.5. By the Corollary in §4.3,

$$\underline{L}_{\underline{\dot{x}}\,\underline{\dot{x}}^*} \cdot \underline{H}_{\underline{y}\,\underline{y}^*} = I_m\ .$$

4.6. To each local Lagrange function $\underline{L}$ there corresponds a local Hamilton function $\underline{H}$. There is the following converse of this proposition:

To any locally defined real function $\underline{H}(\underline{x},\underline{y},t)$ with a non-singular Hessian $H_{\underline{y}\,\underline{y}^*}$ there corresponds a locally defined function $\underline{L}(\underline{x},\underline{\dot{x}},t)$, with a non-singular Hessian $\underline{L}_{\underline{\dot{x}}\,\underline{\dot{x}}^*}$, given by

$$\begin{cases} \underline{\dot{x}} = \underline{H}_{\underline{y}}(\underline{x},\underline{y},t) \\[2ex] \underline{L}(\underline{x},\underline{\dot{x}},t) + H(\underline{x},\underline{y},t) = \underline{y}^*\ \underline{x}\ . \end{cases}$$

In particular, if $\underline{H}$ is the local Hamilton function corresponding to a given local Lagrange function $\underline{L}$, then $\underline{L}$ is the function obtained from $\underline{H}$ by the above proposition.

The assertions follow immediately from Lemma 4.3.

4.7. We now wish to show that to the given manifold M there corresponds a global coordinate-independent Hamilton function H, in a similar way as to M there corresponds a global coordinate-independent Lagrange function L.

First it will be demonstrated that the vectors $\underline{y}$ are representatives of covariant vectors (whereas, as was shown in §3.12, the vectors $\dot{x}$ are contravariant). Assign, to each point x of the m-dimensional analytic M, an m-dimensional vector space $\underset{\bullet}{M}(x)$ the elements of which are covariant vectors. Now, let $\underline{\dot{x}}$ and $\underline{\dot{\bar{x}}}$ be two representatives of the contravariant vector $\dot{x} \in \dot{M}(x)$, and $\underline{y}$ and $\underline{\bar{y}}$ be the corresponding "momenta" given by

$$\underline{y} = \underline{L}_{\underline{\dot{x}}} \quad \text{and} \quad \underline{\bar{y}} = \underline{\bar{L}}_{\underline{\dot{\bar{x}}}} \quad .$$

Hence,

$$\bar{y}_i = \frac{\partial \bar{L}}{\partial \dot{\bar{x}}^i} = \frac{\partial L}{\partial \dot{x}^j} \frac{\partial \dot{x}^j}{\partial \dot{\bar{x}}^i} = y_j \frac{\partial x^j}{\partial \bar{x}^i}$$

since

$$\dot{x}^j = \frac{\partial x^j}{\partial \bar{x}^i} \dot{\bar{x}}^i \quad ,$$

and this implies that $\underline{y}$ and $\underline{\bar{y}}$ are representatives of a covariant vector $y \in \underset{\bullet}{M}(x)$.

The "inner product" of a contravariant vector λ and a covariant vector μ is invariant under admissible coordinate transformations $\underline{x} \to \underline{\bar{x}}$: Let $\underline{\lambda}$ and $\underline{\bar{\lambda}}$ be representatives of λ, and $\underline{\mu}$ and $\underline{\bar{\mu}}$ be representatives of μ. Then,

$$\bar{\lambda}^i \bar{\mu}_i = \lambda^i \mu_i$$

as is shown by a short computation. We shall call this coordinate-independent number $\underline{\lambda}^* \underline{\mu}$ the <u>inner product</u> of the vectors λ and μ, and denote it by $\lambda^*\mu$ or $\mu^*\lambda$. The same applies to vector fields.

4.8. Thus in the equation defining the local Hamilton function H,

$$\underline{L} + \underline{H} = \underline{y}^* \dot{\underline{x}} ,$$

the terms $\underline{L}$ and $\underline{y}^* \dot{\underline{x}}$ have a coordinate-independent meaning, and so has, therefore, the term $\underline{H} = \underline{y}^* \dot{\underline{x}} - \underline{L}$.

4.9. Now let $\underline{L}$ and $\bar{\underline{L}}$ be two local representatives of the Lagrange function L in the same coordinate neighbourhood on the manifold M, and let $\underline{H}$ and $\bar{\underline{H}}$ be the local Hamilton functions corresponding to $\underline{L}$ and $\bar{\underline{L}}$. Then, $\bar{\underline{H}}$ is the transform of $\underline{H}$ under the transformation of coordinates

$$\begin{cases} \bar{\underline{x}} = \bar{\underline{\phi}}(\underline{x},t), \\ \bar{\underline{y}} = \underline{\phi}_{\bar{\underline{x}}^*}{}^*(\bar{\underline{x}},t)\underline{y} \end{cases}$$

where again $\underline{\phi}$ and $\bar{\underline{\phi}}$ are local inverses of each other; that is,

$$\bar{\underline{H}}(\bar{\underline{x}},\bar{\underline{y}},t) = \underline{H}(\phi(\bar{\underline{x}},t), \quad \bar{\underline{\phi}}_{\underline{x}^*}{}^*(\underline{x},t)\bar{\underline{y}},t).$$

Conversely, $\underline{H}$ is the transform of $\bar{\underline{H}}$ under the inverse transformation $(\bar{\underline{x}},\bar{\underline{y}}) \to (\underline{x},\underline{y})$.

Proof:
$$\begin{aligned} \bar{\underline{H}}(\bar{\underline{x}},\bar{\underline{y}}) &= \dot{\bar{\underline{x}}}^*\underline{y} - \bar{\underline{L}}(\bar{\underline{x}},\dot{\bar{\underline{x}}},t) \\ &= \dot{\underline{x}}^*\underline{y} - \underline{L}(\underline{x},\dot{\underline{x}},t) \quad (\S 4.2) \\ &= \underline{H}(\underline{x},\underline{y}). \end{aligned}$$

Thus the following commutative diagram holds:

$$\begin{array}{ccccc} & \underline{L} & \to & \underline{H} & \\ \bar{\underline{\phi}} & \downarrow & & \downarrow & \bar{\underline{\phi}} \\ & \bar{\underline{L}} & \to & \bar{\underline{H}} & \end{array}$$

4.10. A global Hamilton function H can now be defined by $H(x,y,t) = \underline{H}(\underline{x},\underline{y},t)$ for any $x \in M$, $y \in \dot{M}(x)$, $t \in T'$ where $x \to \underline{x}$ is a local system of coordinates and $\underline{y}$ is a representative of the covariant vector $y \in \dot{M}(t)$. Evidently

$$H(x,y,t) + L(x,\dot{x},t) = \dot{x}^*y.$$

4.11. Finally, since the local Hamilton equations

$$\frac{d\underline{x}}{dt} = \underline{H}_{\underline{y}}(\underline{x},\underline{y},t),$$

$$\frac{d\underline{y}}{dt} = -\underline{H}_{\underline{x}}(\underline{x},\underline{y},t)$$

are invariant under local transformations of coordinates, they may be regarded as local representatives of global Hamilton equations

$$\left.\begin{aligned} \frac{dx}{dt} &= H_y(x,y,t), \\ \frac{dy}{dt} &= -H_x(x,y,t), \end{aligned}\right\} \qquad x \in M,\ y \in \dot{M}(x),\ t \in T'.$$

These equations determine motions $t \to x(t)$ on the manifold which are identical with the Lagrange curves determined by the Lagrange function $L = \dot{x}^*y - H$.

If only time-independent transformations of local coordinates are considered, the first equation expresses the equality of two contravariant vectors, and the second equation expresses that of two covariant vectors.

References

Birkhoff (1927), Wintner (1952), Siegel (1956), Whittaker (1961), MacLane (1970), Siegel-Moser (1971).

§5. Canonical Transformations

5.1. In this and the following sections (excepting §5.17) only local questions will be discussed. The points considered will belong to a sufficiently small open neighbourhood of a given point $(\overset{o}{x},\overset{o}{y})$. It then will no longer be strictly necessary to distinguish, in the notation applied, between a point and its coordinate vector - although, of course, the conceptual distinction is to be maintained clearly. Thus the notation can be simplified. Symbols like "x" and "y" will now be used ambiguously: they may denote points, but they may also denote the corresponding vectors. But what is meant is usually unmistakable from the context. So we shall, for example, write "x" instead of the symbol "$\underline{x}$" used thus far.

5.2. By §3.7, any admissible transformation of coordinates transforms Lagrange's equations of motion into equations of the same form. The one-to-one correspondence between Lagrange's and Hamilton's equations (§4.9), therefore, suggests that such a transformation transforms Hamiltonian equations into Hamiltonian equations. These transformations take, for Hamiltonian equations, the fo

$$\begin{cases} \bar{x} = \bar{\phi}(x,\ t), \\ \bar{y} = \bar{\phi}_{\underline{x}^*}(x,y,t)\ y \underset{\text{Def.}}{=} \bar{\psi}(x,y,t). \end{cases}$$

It is natural to ask under which transformations of the more general form

$$\begin{cases} \bar{x} = \bar{\phi}(x,y,t)\ , \\ \bar{y} = \bar{\psi}(x,y,t) \end{cases}$$

the Hamiltonian form of the equations of motion is preserved. Those admissible coordinate transformations which preserve this character for every Hamilton function will be called (local) canonical transformations. We are going to establish necessary and sufficient conditions for them.

5.3. For this purpose it is convenient to introduce the following notation.

Let I_m denote the (m×m) unit matrix,

$$J = \begin{pmatrix} 0 & I_m \\ -I_m & 0 \end{pmatrix} \qquad \text{(Def.)}$$

(where 0 stands for the (m×m) zero matrix),

$$z = \begin{pmatrix} x \\ y \end{pmatrix} \qquad \text{(Def.)}$$

(thus z being a (2m)-column),

$$H(z,t) = H(x,y,t) ,$$
$$H_z(z,t) = \begin{pmatrix} H_x(x,y,t) \\ H_y(x,y,t) \end{pmatrix} . \qquad \text{(Def.)}$$

Then the Hamilton equations

$$\begin{cases} \dfrac{dx}{dt} = H_y(x,y,t) , \\ \dfrac{dy}{dt} = -H_x(x,y,t) \end{cases}$$

are equivalent to the equation

$$\frac{dz}{dt} = J\, H_z(z,t).$$

5.4. Lemma. Let the (n×n) - matrix A and the n-column b be given functions which are defined in some region of the n-dimensional Euclidean number space E^n; let α be an arbitrary real function defined in the same region and α_z be the gradient of α. Then, $A\alpha_z + b$ is a gradient for every α if, and only if, the vector b is a gradient and the matrix A is a constant multiple of the (n×n) unit matrix.

Proof. The "if" part is obvious. We turn to the "only if" part, assuming that $A\alpha_z + b$ is a gradient for each admitted function α.

(1) $\alpha = 0$ is an admitted function: hence b is a gradient.

(2) $A\alpha_z$ is a gradient. (Hypothesis) (1)

(3) Let $A = \{A_i^j\}_{i,j=1}^{n}$, and $z = \{z^k\}_{k=1}^{n}$. Then,

$$\frac{\partial}{\partial z^j}\left(A_i^k \frac{\partial\alpha}{\partial z^k}\right) = \frac{\partial}{\partial z^i}\left(A_j^k \frac{\partial\alpha}{\partial z^k}\right) \qquad (2)$$

(4) $\alpha(z) = z^\ell$, $\ell=1,\ldots,n$, is an admitted function:

hence

$$\frac{\partial}{\partial z^j} A_i^\ell = \frac{\partial}{\partial z^i} A_j^\ell \ . \qquad (3)$$

(5) For any admitted α ,

$$A_i^k \frac{\partial^2\alpha}{\partial z^j \partial z^k} = A_j^k \frac{\partial^2\alpha}{\partial z^i \partial z^k} \ . \qquad (3)(4)$$

(6) $\alpha(z) = \frac{1}{2} z^1 z^1$ is an admitted function:

hence,

$$A_i^1 \delta_{1j} = A_j^1 \delta_{1i}$$

(where $\{\delta_{ij}\}_{i,j=1}^{n}$ is the unit matrix).

(7) $A_i^1 = 0$ if $i \neq 1$. (6)

(8) Similarly,

$A_i^j = 0$ if $i \neq j$. (6)(7)

(9) The matrix A is diagonal. (8)

(10) $\frac{\partial}{\partial z^j} A_1^1 = \frac{\partial}{\partial z^1} A_j^1$. (4)

(11) $\frac{\partial}{\partial z^j} A_1^1 = 0$ if $j \neq 1$. (10)

(12) A_1^1 is a function of z^1 alone. (11)

(13) Similarly, A_2^2 is a function z^2 alone, etc. (12)

(14) $\frac{\partial}{\partial z^2} \left(A_1^1 \frac{\partial \alpha}{\partial z^1} \right) = \frac{\partial}{\partial z^1} \left(A_2^2 \frac{\partial \alpha}{\partial z^2} \right)$. (3)(12)(13)

(15) $A_1^1 \frac{\partial^2 \alpha}{\partial z^1 \partial z^2} = A_2^2 \frac{\partial^2 \alpha}{\partial z^1 \partial z^2}$

for all admitted functions α .

(16) $A_1^1 (z^1) = A_2^2 (z^2) = \cdots = A_n^n (z^n)$. (15)

(17) A_1^1 , A_2^2 ,$\cdots$, A_n^n are equal to some fixed real number, as has been asserted. (16)

<u>5.5.</u> <u>Let</u>

$$\bar{z} = \bar{\chi} (z,t)$$

<u>define a canonical transformation of local coordinates, and</u> $z = \chi(\bar{z},t)$,

for any fixed t, define its local inverse. Then,

$$\begin{cases} \bar{\chi}_{z*} J \bar{\chi}_{z*}^{*} = \mu J , \\ \bar{\chi}_{z*}^{*} J \bar{\chi}_{z*} = \mu J \end{cases}$$

for some non-vanishing real number μ, and $J^*\bar{\chi}_t$ $(\chi(\bar{z},t),t)$ is the gradient of some function $\bar{R}$ of $(\bar{z},t)$ with respect to $\bar{z}$. The subscript "t" again indicates derivatives with respect to the time variable t. The arbitrary local Hamilton function $(z,t) \to H(z,t)$ is transformed by $\bar{\chi}$ into the local Hamilton function $\bar{K}$, given by

$$\bar{K}(\bar{z},t) = \mu\bar{H}(\bar{z},t) + \bar{R}(\bar{z},t)$$

where

$$\bar{H}(\bar{z},t) \underset{\text{Def.}}{=} H(\chi(\bar{z},t), t)$$

and

$$\bar{R}_{\bar{z}}(\bar{z},t) = J^* \bar{\chi}_t(\chi(\bar{z},t), t).$$

The real number μ is called the multiplier of the transformation $\bar{\chi}$.

Proof. First, suppose that $\bar{\chi}$ is canonical, and let $t \to z(t)$ be any solution of the Hamilton equation corresponding to the Hamilton function H. Then,

(1) $\dfrac{dz}{dt} = JH_z\ (z,t)$.

(2) Let $\bar{z}(t) \underset{\text{Def.}}{=} \bar{\chi}(z(t),t)$. Then,

$$\frac{d\bar{z}}{dt} = \bar{\chi}_t + \bar{\chi}_{z*} \frac{dz}{dt} \qquad \text{(Hyp.)}$$

$$= \bar{\chi}_t + \bar{\chi}_{z*} JH_z(z,t) \qquad (1)$$

$$\doteq \bar{\chi}_t + \bar{\chi}_{z*} J\bar{\chi}_{z*}^{*}\bar{H}_{\bar{z}}\ (\bar{z},t)$$

(3) $\frac{d\bar{z}}{dt} = J \cdot (J^* \bar{\chi}_t + J^* \bar{\chi}_{z^*} J \bar{\chi}_{z^*}{}^* \bar{H}_{\bar{z}})$ (2)

since

$$J^* = -J = J^{-1} .$$

(4) $\frac{d\bar{z}}{dt} = J\bar{K}_{\bar{z}} (\bar{z},t)$

for some real function $\bar{K}$ of $(\bar{z},t)$. (Hypothesis)

(5) $J^* \bar{\chi}_t + J^* \bar{\chi}_{z^*} J \bar{\chi}_{z^*}{}^* \bar{H}_{\bar{z}} = \bar{K}_{\bar{z}} (\bar{z},t)$

for every solution $t \to \bar{z}(t)$ and, since its initial value is arbitrary, for every $(\bar{z},t)$ where $\bar{z}$ is an independent variable. (3)(4)

(6) That is,

$$J^* \bar{\chi}_t(\chi(\bar{z},t),t) + J^* \bar{\chi}_{z^*}(\chi(\bar{z},t),t) J \bar{\chi}_{z^*}{}^* (\chi(\bar{z},t),t) \cdot \bar{H}_{\bar{z}}(\bar{z},t)$$
$$= \bar{K}_{\bar{z}} (\bar{z},t)$$

identically in $(\bar{z},t)$.

(7) $J^* \bar{\chi}_t(\chi(\bar{z},t),t)$ is the gradient of some function $\bar{R}$ of $(\bar{z},t)$ with respect to $\bar{z}$. (6)(§ 5.4)

(8) $J^* \bar{\chi}_{z^*} J \bar{\chi}_{z^*}{}^* = \mu(t) I_{2m}$ for some function μ of t. (6)(§ 5.4)

(9) $\bar{R}_{\bar{z}} + \mu \bar{H}_{\bar{z}} = \bar{K}_{\bar{z}}$. (6)(2)(8)

(10) $\mu(t) \neq 0$ for all admitted t; for,

$$\mu^{2m} = \det(\mu I_{2m}) = \det(J^* \bar{\chi}_{z^*} J \bar{\chi}_{z^*}{}^*) \qquad (8)$$

$$= (\det J)^2 \cdot (\det \bar{\chi}_{z^*})^2$$

$$= (\det \bar{\chi}_{z^*})^2$$

$$> 0.$$

It remains to show that the function μ is a constant:

(11) Since $J^* \bar{\chi}_t$ is, as a function of $\bar{z}$, a gradient with respect to $\bar{z}$,

$$(J^* \bar{\chi}_t)_{\bar{z}^*} = (J^* \bar{\chi}_t)_{\bar{z}^*}{}^* . \qquad (7)$$

(12) $$J^* \bar{\chi}_{tz^*} \chi_{\bar{z}^*} = (J^* \bar{\chi}_{tz^*} \chi_{\bar{z}^*})^* = \chi_{\bar{z}^*}{}^* \bar{\chi}_{tz^*}{}^* J. \qquad (11)$$

(13) Multiply (12) with $\bar{\chi}_{z^*}{}^*$ from the left, and with $\bar{\chi}_{z^*}$ from the right: thus,

$$\bar{\chi}_{z^*}{}^* J^* \bar{\chi}_{tz^*} = \bar{\chi}_{tz^*}{}^* J \bar{\chi}_{z^*}$$

since

$$\bar{\chi}_{z^*}{}^* \chi_{\bar{z}^*}{}^* = (\chi_{\bar{z}^*} \cdot \bar{\chi}_{z^*})^* = I^*_{2m} = I_{2m} .$$

(14) $$\bar{\chi}_{tz^*}{}^* J \bar{\chi}_{z^*} + \bar{\chi}_{z^*}{}^* J \bar{\chi}_{tz^*} = 0 \qquad (13)$$

since $J^* = -J$.

(15) $$(\bar{\chi}_{z^*}{}^* J \bar{\chi}_{z^*})_t = 0. \qquad (14)$$

(15) $$\bar{\chi}_{z^*}{}^* J \bar{\chi}_{z^*} = C(z)$$

for some matrix function C of z. (15)

(16) $\mu \cdot \bar{\chi}_{z^*} J \bar{\chi}_{z^*}^* = J.$ (8)

(17) $(\mu \bar{\chi}_{z^*} J \bar{\chi}_{z^*}^*)(\bar{\chi}_{z^*}^* J \bar{\chi}_{z^*}) = JC(z)$ (15)(16)

(18) $-\mu I_{2m} = J\, C(z)$ (17)

(19) $\left.\begin{aligned} \mu(t) &= \text{const.}, \\ C(z) &= \mu J. \end{aligned}\right\}$ (18)

(20) $\bar{\chi}_{z^*}^* J \bar{\chi}_{z^*} = \mu J.$ (19)

Thus all the assertions of the proposition have been proved.

5.6. Conversely, *let* $\bar{\chi}$ *be a (possibly time-dependent) transformation of local coordinates such that, for some non-vanishing real number* μ,

$$\bar{\chi}_{z^*} J \bar{\chi}_{z^*}^* = \mu J$$

and $J^* \bar{\chi}_t$ *is the gradient of some function* $\bar{R}$ *of* $(\bar{z}, t)$ *with respect to* $\bar{z}$. *Then* $\bar{\chi}$ *is canonical and transforms the arbitrary local Hamilton function* $(z,t) \to H(z,t)$ *into the local Hamilton function*

$$\bar{K} \underset{\text{Def.}}{=} \mu \bar{H} + \bar{R}\ .$$

(The notation is the same as in §5.5.)

Proof. Let $t \to z(t)$ be any local solution of the Hamilton equation

$$\frac{dz}{dt} = JH_z(z,t),$$

and

$$\bar{z}(t) \underset{\text{Def.}}{=} \bar{\chi}(z(t),t).$$

Abbreviating z(t) and $\overline{z}(t)$ by z and $\overline{z}$, we obtain

$$\frac{d\overline{z}}{dt} = \frac{d}{dt}\overline{\chi}(z,t)$$

$$= \overline{\chi}_{z*}\frac{dz}{dt} + \overline{\chi}_t$$

$$= \overline{\chi}_{z*}JH_z + \overline{\chi}_t$$

$$= \overline{\chi}_{z*}J\,\overline{\chi}_{z*}{}^*\,\overline{H}_{\overline{z}} + J\overline{R}_{\overline{z}} \qquad \text{(Hyp.)}$$

$$= \mu J\overline{H}_{\overline{z}*} + J\overline{R}_{\overline{z}} \qquad \text{(Hyp.)}$$

$$= J\overline{K}_{\overline{z}} \qquad \text{(Hyp.)},$$

which proves the assertion.

Note that in this converse of Proposition 5.5 it is not supposed that

$$\overline{\chi}_{z*}{}^*J\,\overline{\chi}_{z*} = \mu J.$$

This relation now follows by an application of Proposition 5.5. It will be shown that, in Proposition 5.6, it can be substituted for the assumption that

$$\overline{\chi}_{z*}\,J\,\overline{\chi}_{z*}{}^* = \mu J.$$

The proof is an immediate consequence of the following proposition.

5.7. If $\overline{\chi}$ is a canonical transformation of multiplier μ, then its inverse is a canonical transformation of multiplier 1/μ .

Proof. By Proposition 5.6 it has to be shown that

$$\chi_{\overline{z}*}\,J\,\chi_{\overline{z}*}{}^* = \frac{1}{\mu}J,$$

and that $J^*\chi_t(\overline{\chi},t)$ is the gradient of some function of (z,t) with respect to z

The first assertion follows immediately from the fact that, by Proposition 5.5,

$$\bar{\chi}_{z*} J \bar{\chi}_{z*}^{*} = \mu J.$$

The second one is obtained by a differentiation of the identity

$$\chi(\bar{\chi}(z,t),t) = z$$

with respect to t. Thus,

$$\begin{aligned} \chi_t(\bar{\chi},t) &= - \chi_{\bar{z}*} \bar{\chi}_t \\ &= -\chi_{\bar{z}*} J\bar{R}_{\bar{z}} \\ &= - \chi_{\bar{z}*} J\chi_{\bar{z}*}^{*} R_z \end{aligned}$$

where $\bar{R}$ has the same meaning as in §5.5 and §5.6, and

$$R(z,t) \underset{\text{Def.}}{=} \bar{R}(\bar{\chi}(z,t),t).$$

Hence,

$$\begin{aligned} J^*\chi_t(\bar{\chi},t) &= -J^* \chi_{\bar{z}*} J\chi_{\bar{z}*}^{*} R_z \\ &= - \frac{1}{\mu} R_z \\ &= \left(- \frac{1}{\mu} R\right)_z . \end{aligned}$$

This completes the proof of the proposition.

5.8. Corollary: If the canonical transformation $\bar{\chi}$ transforms the Hamilton function H into the Hamilton function $\bar{K}$, then the inverse canonical transformation χ transforms $\bar{K}$ into H.

5.9. We shall now prove the remark made at the end of §5.6., i.e., the second converse of Proposition 5.5:

Let $\bar{\chi}$ be a (possibly time-dependent) transformation of local coordinates such that

$$\bar{\chi}_{z*}^{*} J \bar{\chi}_{z*} = \mu J$$

for some non-vanishing real μ and $J^*\bar{\chi}_t$ is the gradient of some function $\bar{R}$ of $(\bar{z},t)$ with respect to $\bar{z}$. Then $\bar{\chi}$ is canonical and transforms the arbitrary local Hamilton function H into the local Hamilton function

$$\bar{K} \underset{\text{Def.}}{=} \mu\bar{H} + \bar{R}.$$

Proof. Let χ be the local inverse of $\bar{\chi}$. By hypothesis,

$$\chi_{\bar{z}*} J \chi_{\bar{z}*}^{*} = \frac{1}{\mu} J.$$

By the proof of Proposition 5.7, $J^*\chi_t(\bar{\chi},t)$ is a gradient with respect to z. Hence, by Proposition 5.6, χ is canonical, and so is therefore, by Proposition 5.7, the inverse $\bar{\chi}$ of χ. Clearly $\bar{\chi}$ transforms H into $\bar{K}$.

5.10. From Proposition 5.5., 5.6., and 5.9 it follows that:

if $\bar{\chi}$ is a time-independent transformation of local coordinates, then either identity,

$$\bar{\chi}_{z*} J \bar{\chi}_{z*}^{*} = \mu J$$

or

$$\bar{\chi}_{z*}^{*} J \bar{\chi}_{z*} = \mu J$$

for some non-vanishing real μ, is a necessary and sufficient condition for $\bar{\chi}$ being canonical.

The conditions

$$\chi_{\bar{z}*} J \chi_{\bar{z}*}^{*} = \frac{1}{\mu} J$$

and

$$\chi_{\bar{z}*}^{*} J \chi_{\bar{z}*} = \frac{1}{\mu} J,$$

where χ denotes the local inverse of $\bar{\chi}$ for fixed t, are equivalent to those given above.

5.11. In particular,

<u>if</u> $\bar{\chi}$ <u>is a time-dependent canonical transformation, and</u> τ <u>is some fixed value which can be taken by the time variable t, then the time-independent transformation</u> $z \to \bar{\chi}(z,\tau)$ <u>is canonical</u>.

5.12. Because of their importance, the conditions of §5.10 will now be repeated in a slightly more explicit form:

<u>The transformation of local coordinates</u>

$$\begin{cases} \bar{x} = \bar{\phi}(x,y), \\ \bar{y} = \bar{\psi}(x,y) \end{cases}$$

<u>is canonical, with multiplier</u> μ, <u>if and only if</u>

$$\begin{cases} \bar{\phi}_{x^*} \bar{\phi}_{y^*}{}^* - \bar{\phi}_{y^*} \bar{\phi}_{x^*}{}^* = 0, \\ \bar{\phi}_{x^*} \bar{\psi}_{y^*}{}^* - \bar{\phi}_{y^*} \bar{\psi}_{x^*}{}^* = \mu I_m, \\ \bar{\psi}_{x^*} \bar{\psi}_{y^*}{}^* - \bar{\psi}_{y^*} \bar{\psi}_{x^*}{}^* = 0 . \end{cases}$$

Written in their components, these matrix equations read

$$\left.\begin{aligned} \frac{\partial \bar{\phi}^i}{\partial x^j} \frac{\partial \bar{\phi}^k}{\partial y_j} - \frac{\partial \bar{\phi}^i}{\partial y_j} \frac{\partial \bar{\phi}^k}{\partial x^j} &= 0, \\ \frac{\partial \bar{\phi}^i}{\partial x^j} \frac{\partial \bar{\psi}_k}{\partial y_j} - \frac{\partial \bar{\phi}^i}{\partial y_j} \frac{\partial \bar{\psi}_k}{\partial x^j} &= \mu \delta^i_k , \\ \frac{\partial \bar{\psi}_i}{\partial x^j} \frac{\partial \bar{\psi}_k}{\partial y_j} - \frac{\partial \bar{\psi}_i}{\partial y_j} \frac{\partial \bar{\psi}_k}{\partial x^j} &= 0, \end{aligned}\right\} \quad \begin{aligned} &i=1,\cdots,m, \\ &k=1,\cdots,m, \end{aligned}$$

where $\{\delta^i_k\} = I_m$. The left-hand sides of these equations, i.e., the elements of the matrix $\overline{\chi}_{z*}J\,\overline{\chi}_{z*}{}^*$, are called Poisson's bracket-expressions.

5.13. Similarly, the transformation of local coordinates

$$\begin{cases} \overline{x} = \overline{\phi}(x,y), \\ \overline{y} = \overline{\psi}(x,y) \end{cases}$$

is canonical, with multiplier μ, if and only if

$$\begin{cases} \overline{\phi}_{x*}{}^* \overline{\psi}_{x*} - \overline{\psi}_{x*}{}^* \overline{\phi}_{x*} = 0, \\ \overline{\phi}_{x*}{}^* \overline{\psi}_{y*} - \overline{\psi}_{x*}{}^* \overline{\phi}_{y*} = \mu I_m, \\ \overline{\phi}_{y*}{}^* \overline{\psi}_{y*} - \overline{\psi}_{y*}{}^* \overline{\phi}_{y*} = 0. \end{cases}$$

Written in components these matrix equations read

$$\begin{cases} \dfrac{\partial\overline{\phi}^j}{\partial x^i}\dfrac{\partial\overline{\psi}_j}{\partial x^k} - \dfrac{\partial\overline{\psi}_j}{\partial x^i}\dfrac{\partial\overline{\phi}^j}{\partial x^k} = 0, \\ \dfrac{\partial\overline{\phi}^j}{\partial x^i}\dfrac{\partial\overline{\psi}_j}{\partial y_k} - \dfrac{\partial\overline{\psi}_j}{\partial x^i}\dfrac{\partial\overline{\phi}^j}{\partial y_k} = \mu\delta^i{}_k, \\ \dfrac{\partial\overline{\phi}^j}{\partial y_i}\dfrac{\partial\overline{\psi}_j}{\partial y_k} - \dfrac{\partial\overline{\psi}_j}{\partial y_i}\dfrac{\partial\overline{\phi}^j}{\partial y_k} = 0. \end{cases}$$

The left-hand sides of these equations, i.e., the elements of the matrix $\overline{\chi}_{z*}{}^*J\,\overline{\chi}_{z*}$, are called Lagrange's bracket-expressions.

5.14. By Proposition 5.12 or 5.13 it is verified that the transformation of local coordinates

$$\begin{cases} \overline{x} = x, \\ \overline{y} = \mu y, \; \mu \neq 0, \end{cases}$$

is canonical with multiplier μ. This result will be used in the following.

5.15. The canonical transformations of local coordinates defined in appropriate neighborhoods of a given point $(\overset{\circ}{x},\overset{\circ}{y})$ form a group. The set of the time-independent canonical transformations is a subgroup of this group.

For, the composition of canonical transformations is canonical by their definition (§5.2); the identity mapping is evidently canonical; and every canonical mapping has a local inverse (§5.7).

5.16. The local canonical transformations of multiplier 1 form a normal subgroup of the group of all local canonical transformations.

For, from Proposition 5.9 it follows that the mapping $\overline{\chi} \to \mu$ which assigns, to any canonical transformation $\overline{\chi}$, its multiplier μ is a homomorphism. Its kernel consists of the transformations having multiplier 1. Hence the assertion.

Canonical transformations of multiplier 1 are called completely canonical.

The set of all transformations of any given multiplier μ is a coset of the normal subgroup of completely canonical transformations. Therefore it is obtained by composing the members of the normal subgroup with any fixed transformation of multiplier μ, e.g. with the transformation considered in §5.14. Because this transformation is so trivial, i.e., because it does not introduce any essential change of the formal or physical aspects of the Hamilton equations of motion, we shall, from now on, restrict our discussion to completely canonical transformations.

5.17. At a first glance it may appear tempting to interpret the invariance of the Hamilton equation

$$\frac{dz}{dt} = JH_z(z,t)$$

under canonical transformations as the invariance of an equation which expresses the equality of vector fields. A second glance, however, shows that this interpretation, in this form, cannot be correct since the Hamilton equations are not invariant under arbitrary transformations of local coordinates. But it can be made correct if the notions of vector and tensor fields are generalized. In §3.9, §3.13, and §3.20, those notions were based on certain equivalence relations, and these, in turn, were based on the group of all transformations of local coordinates. Thus, for example, the equivalence relation defining contravariant vector fields, reads

$$\bar{\lambda}^i = \frac{\partial \bar{x}^i}{\partial x^j} \lambda^i.$$

The concept of a contravariant vector field can now be generalized by restricting the admissible transformations $x \to \bar{x}$ to some given subgroup of the group of all coordinate transformations; for instance, to the subgroup of the completely canonical transformations. Thus we are lead to the following definitions:

Let M be a (2m)-dimensional analytic manifold on which systems of time-independent local coordinates are introduced in such a way that the coordinate transformations between any two "overlapping" systems of local coordinates are completely canonical. Only time-independent completely canonical transformations of coordinates shall be considered as admissible. Then let us say the manifold M is <u>completely canonical</u>. On M we can now define <u>vector and tensor fields relative to the group of the time-independent completely canonical transformations of local coordinates</u>, in the manner indicated above. The necessary and sufficient condition for a (time-independent) coordinate transformation to be completely canonical,

$$\bar{\chi}_{z*}{}^{*}J\bar{\chi}_{z*} = J,$$

shows that the matrix J is the representative of a contravariant tensor field of the second order whose representative in <u>any</u> admissible system of coordinates is J.

Suppose further that on the completely canonical manifold M a real function H is given which may depend on the time variable t. It determines, for any fixed value of t, a covariant vector field H_z on M whose representative in any local coordinate system is the gradient of the representative of H with respect to the local coordinate vector. Thus the Hamilton equation

$$\frac{dz}{dt} = JH_z(z,t),$$

where z denotes a point on the manifold, expresses the equality of the contravariant vector fields dz/dt and JH_z. The curves which solve any local representative of this Hamilton equation on the manifold are invariant under admissible transformations of local transformations. Thus the Hamilton equations on the completely canonical manifold define solution curves on the manifold.

References

Wintner (1952), Siegel (1956), Courant-Hilbert (1962), Gelfand-Fomin (1963), Siegel-Moser (1971). For the elements of group theory see, for example, Van der Waerden (1964), Herstein (1964), or Polites (1968).

§6. Time-Dependent Completely Canonical Transformations

6.1. In this section a survey over and the principal application of time-dependent completely canonical transformations will be given. The discussion will be local.

Let

$$\bar{z} = \bar{\chi}(z,t),$$

with

$$\bar{\chi}_t \neq 0,$$

be a time-dependent completely canonical transformation of local coordinates. Then $\bar{\chi}$ *is the local solution of a local Hamilton equation which satisfies, for arbitrary* z, *the initial condition*

$$z = \bar{\chi}(z,0).$$

(It is assumed that 0 belongs to the time interval T' considered.)

Proof. By Proposition 5.5,

$$J^{-1}\,\bar{\chi}_t(\chi(\bar{z},t),t) = \bar{R}_{\bar{z}}(\bar{z},t)$$

for some real function $\bar{R}$, where χ again denotes, for any fixed value of t, the inverse of $\bar{\chi}$. Hence,

$$\bar{\chi}_t(z,t) = J\bar{R}_{\bar{z}}(\bar{\chi}(z,t),t),$$

which implies the assertion.

6.2. This proposition can be given the form:

If the time-dependent completely canonical transformation $\bar{\chi}$, *with* $\bar{\chi}_t \neq 0$, *satisfies the "distinguished" initial condition"*

$$\bar{\chi}(z,0) = z \quad \textit{for all admitted } z,$$

then $\bar{\chi}$ *is the general local solution of the local Hamilton equation*

$$\frac{d\bar{z}}{dt} = J\bar{R}_{\bar{z}}(\bar{z},t),$$

where $\bar{R}$ *denotes some real function of* $(\bar{z},t)$.

6.3. This proposition has the following converse:

Let $\bar{R}$ *be any real function of* $(\bar{z},t)$, *and*

$$\bar{z} = \bar{\chi}(z,t),$$

be the general solution of the Hamilton equation

$$\frac{d\bar{z}}{dt} = J\bar{R}_{\bar{z}}(\bar{z},t),$$

satisfying the initial condition

$$\bar{\chi}(z,0) = z \quad \text{for all admitted } z.$$

Then the function $\bar{\chi}$, considered as a time-dependent transformation $(z,t) \to \bar{z}$ *of local coordinates, is completely canonical.*

Proof. $J^* \bar{\chi}_t(\chi,t)$ is the gradient of the real function $\bar{R}$. Hence, by Proposition 5.6, the second condition for $\bar{\chi}$ being canonical is fulfilled. It remains to show that the first condition,

$$\bar{\chi}_{z*} J\bar{\chi}_{z*}{}^* = J,$$

is also satisfied. For this end, define

$$M \underset{\text{Def.}}{=} \bar{\chi}_{z*} J\bar{\chi}_{z*}{}^*.$$

It now suffices to prove that

$$M = J.$$

By the definition of M,

$$M_t = \bar{\chi}_{z*t} J\bar{\chi}_{z*}{}^* + \bar{\chi}_{z*} J\bar{\chi}_{z*t}{}^*$$

and, from the given Hamilton equation,

$$\bar{\chi}_{z*t} = J\bar{R}_{\bar{z}\bar{z}*}\bar{\chi}_{z*}.$$

Thus,

$$M_t = J\bar{R}_{\bar{z}\bar{z}*}\bar{\chi}_{z*} J\bar{\chi}_{z*}{}^* + \bar{\chi}_{z*} J\bar{\chi}_{z*}{}^*\bar{R}_{\bar{z}\bar{z}*}{}^* J^*,$$

that is,

$$M_t = J\bar{R}_{\bar{z}\bar{z}*} M - M\bar{R}_{\bar{z}\bar{z}*} J.$$

The constant function J satisfies both this differential equation for M and the initial condition

$$M(\bar{z},0) = J \ .$$

Hence,

$$M = J$$

as has been asserted.

Corollary. The condition of the proposition that the transformation $\bar{\chi}$ satisfies the distinguished initial condition

$$\bar{\chi}(z,0) = z$$

or, equivalently,

$$\bar{\chi}_{z*}(z,0) = I_{2m},$$

can be replaced by the more general condition that the time-independent transformation $\bar{z} = \bar{\chi}(z,0)$ is completely canonical, or

$$\bar{\chi}_{z*}(z,0)J\bar{\chi}_{z*}{}^*(z,0) = J.$$

6.4. Now the conditions of Proposition 5.6 for a time-dependent transformation of local coordinates to be completely canonical can be relaxed:

A time-dependent transformation $\bar{\chi}$ of local coordinates, with $\bar{\chi}_t \neq 0$ and

$$\bar{\chi}(z,0) = z \quad \text{for all admitted } z,$$

is completely canonical if, and only if, $J^*\bar{\chi}_t(\chi,t)$ is the gradient of some real function of $(\bar{z},t)$ with respect to $\bar{z}$.

The "only if" part of the proposition is contained in Proposition 5.5. As to the "if" part, the transformation $\bar{\chi}$ satisfies, by §6.1, a Hamilton equation with non-vanishing right-hand side. Proposition 6.3 now implies the assertion.

6.5. Liouville's Theorem, of importance in Statistical Mechanics, is a corollary to Proposition 6.3. It reads:

Let

$$\bar{z} = \bar{\chi}(z,t),$$

with

$$\bar{\chi}\ (z,0) = z \quad \text{for all admitted } z,$$

be the general local solution of a Hamilton equation. Then,

$$\det \bar{\chi}_{z*} = 1.$$

Proof. (1) $\bar{\chi}$ is a completely canonical transformation. (§6.3)

(2) $\bar{\chi}_{z*} J \bar{\chi}_{z*}{}^{*} = J$. (§5.5, §5.16)

(3) $\det \bar{\chi}_{z*} \cdot \det J \cdot \det \bar{\chi}_{z*} = \det J$. (2)

(4) $\det \bar{\chi}_{z*} = \pm 1$. (3)

(5) $\det \bar{\chi}_{z*}(z,0) = \det I_{2m} = 1$. (Hyp.)

(6) $\det \bar{\chi}_{z*}$ is a continuous function of (z,t).

(7) $\det \bar{\chi}_{z*} = 1$. (4)-(6)

6.6. We now ask: What is the relation between the general completely canonical transformation and those with distingished initial condition $\bar{\chi}(z,0) = z$?

The following answer is an immediate consequence of the appropriate definitions:

The completely canonical transformations with distinguished initial condition, defined in the appropriate neighbourhoods of a given point, form a normal subgroup of the group of the completely canonical transformations in those neighbourhoods. This normal subgroup is the kernel of the homomorphism $\bar{\chi} \to \bar{\chi}(\cdot,0)$ where $\bar{\chi}$ denotes any (local) completely canonical transformation. Any such transformation is the composition of a completely canonical transformation with distinguished initial condition, and a time-independent completely canonical transformation.

Propositions 6.2 and 6.3 yield a survey over all completely canonical transformations with distinguished initial conditions. Thus, by the last result, a survey over all completely canonical transformations is obtained when all time-independent canonical transformations are known. In the next chapter, a certain class of these transformations will be constructed. First, however, Propositions 6.2 and 6.3 will be used for establishing a main result in the theory of perturbations.

6.7. Let

$$\bar{z} = \bar{\chi}(z,t),$$

with

$$\bar{\chi}(z,0) = \bar{z} \quad \text{for all admitted } z,$$

be the general local solution of the Hamilton equation

$$\frac{d\bar{z}}{dt} = J\bar{H}_{\bar{z}}(\bar{z},t),$$

and

$$z = \chi(\bar{z},t)$$

be, for any fixed value of t, the local inverse of the transformation $\bar{\chi}$. Then the completely canonical transformation χ transforms any Hamilton equation of the form

$$\frac{d\bar{z}}{dt} = J\cdot(\bar{H}_{\bar{z}}(\bar{z},t) + \bar{P}_{\bar{z}}(\bar{z},t))$$

into the Hamilton equation

$$\frac{dz}{dt} = J\cdot P_z(z,t)$$

<u>where</u>

$$P(z,t) \underset{\text{Def.}}{=} \overline{P}\,(\overline{\chi}(z,t),t).$$

The transformation χ transforms the equation with the Hamilton function $\overline{H}$ "to rest" (the coordinates of the fixed "starting point" z are taken as the "coordinates of a motion"). Further, the transformation is linear, in the sense that the transform of a sum equals the sum of the transforms. And, finally, since $\overline{\chi}$ is completely canonical (by §6.3), χ is also completely canonical (§5.7). All this makes us expect that the canonical transformation χ, applied to a system with the Hamilton function $\overline{H} + \overline{P}$, will "transform the term $\overline{H}$ away."

<u>Proof</u>. The Hamilton equation

$$\frac{d\overline{z}}{dt} = J\cdot\overline{H}_{\overline{z}}(\overline{z},t)$$

is by the completely canonical transformation

$$z = \chi(\overline{z},t)$$

transformed into the Hamilton equation

$$\frac{dz}{dt} = J\cdot(H + R)_z$$

where

$$H(z,t) \underset{\text{Def.}}{=} \overline{H}(\overline{\chi}(z,t),t)$$

and

$$JR_z(z,t) = \chi_t(\overline{\chi}(z,t),t).$$

(Cf. §5.5.) The general solution of this equation is the transform of the solution $\overline{\chi}$ of the equation $d\overline{z}/dt = J\overline{H}_{\overline{z}}$, i.e., the function $\chi(\overline{\chi}(z,t),t)$, which is identically equal to z. Hence,

$$J\cdot(H + R)_z = \frac{d}{dt}\chi(\overline{\chi}(z,t),t) = 0.$$

Now the transformation χ transforms the equation

$$\frac{d\overline{z}}{dt} = J\cdot(\overline{H} + \overline{P})_{\overline{z}}$$

into

$$\frac{dz}{dt} = J\cdot(H + P + R)_z$$

and, therefore, into

$$\frac{dz}{dt} = J\cdot P_z(z,t)$$

- as has been asserted.

§7. Time-Independent Completely Canonical Transformations: Generating Functions

7.1. Proposition 5.11 yields a method of constructing time-independent completely canonical transformations. However, this method is, in general, impracticable. In the following another method will be expounded by which such transformations can be constructed effectively.

Let, as before, the equations

$$\begin{cases} \overline{x} = \overline{\phi}(x,y), \\ \overline{y} = \overline{\psi}(x,y) \end{cases}$$

describe a transformation $\begin{pmatrix} \overline{\phi} \\ \overline{\psi} \end{pmatrix}$ of local coordinates. Then, by Proposition 5.12 and 5.13,

the transformation $\begin{pmatrix} \overline{\phi} \\ \overline{\psi} \end{pmatrix}$ is completely canonical

$$\Longleftrightarrow \begin{cases} \bar\phi_{x^*} \bar\phi_{y^*}{}^* - \bar\phi_{y^*} \bar\phi_{x^*}{}^* = 0, \\ \bar\phi_{x^*} \bar\psi_{y^*}{}^* - \bar\phi_{y^*} \bar\psi_{x}{}^* = I_m, \\ \bar\psi_{x^*} \bar\psi_{y^*}{}^* - \bar\psi_{y^*} \bar\psi_{x^*}{}^* = 0 \end{cases}$$

$$\Longleftrightarrow \begin{cases} \bar\phi_{x^*}{}^* \bar\psi_{x^*} - \bar\psi_{x^*}{}^* \bar\phi_{x^*} = 0, \\ \bar\phi_{x^*}{}^* \bar\psi_{y^*} - \bar\psi_{x^*}{}^* \bar\phi_{y^*} = I_m, \\ \bar\phi_{y^*}{}^* \bar\psi_{y^*} - \bar\psi_{y^*}{}^* \bar\phi_{y^*} = 0. \end{cases}$$

The symbol "$\Longleftrightarrow$" denotes mutual implication.

7.2. The transformation $\begin{pmatrix} \bar\phi \\ \bar\psi \end{pmatrix}$ is completely canonical $\Longleftrightarrow$ the differential form

$$x^*dy - \bar\phi^*(x,y)d\bar\psi(x,y),$$

abbreviated by

$$x^*dy - \bar x^* d\bar y,$$

is the exact differential of some real-valued function of (x,y).

Proof. Make, in the differential form $x^*dy - \bar x^* d\bar y$, the substitutions

$$\begin{cases} \bar x = \bar\phi(x,y), \\ d\bar y = \bar\psi_{x^*}(x,y)dx + \bar\psi_{y^*}(x,y)dy. \end{cases}$$

Thus the form

$$\begin{pmatrix} -\bar\psi_{x^*}{}^* \bar\phi \\ x \;\; -\bar\psi_{y^*}{}^* \bar\phi \end{pmatrix}^* \begin{pmatrix} dx \\ dy \end{pmatrix}$$

is obtained. A differential form $Z^*(z)\,dz$ is an exact differential if, and only if, the matrix $Z_{z^*}(z)$ is symmetric. A straight-forward computation shows that: $x^*dy - \bar{x}^*d\bar{y}$ is the exact differential of a function of (x,y)

$$\Longleftrightarrow \begin{pmatrix} (-\bar{\psi}_{x^*}{}^*\bar{\phi})_{x^*}, & (-\bar{\psi}_{x^*}{}^*\bar{\phi})_{y^*} \\ (x-\bar{\psi}_{y^*}{}^*\bar{\phi})_{x^*}, & (x-\bar{\psi}_{y^*}{}^*\bar{\phi})_{y^*} \end{pmatrix} \text{ is symmetric}$$

$$\Longleftrightarrow \begin{cases} \bar{\phi}_{x^*}{}^*\bar{\psi}_{x^*} - \bar{\psi}_{x^*}{}^*\bar{\phi}_{x^*} = 0, \\ \bar{\phi}_{x^*}{}^*\bar{\psi}_{y^*} - \bar{\psi}_{x^*}{}^*\bar{\phi}_{y^*} = I_m, \\ \bar{\phi}_{y^*}{}^*\bar{\psi}_{y^*} - \bar{\psi}_{y^*}{}^*\bar{\phi}_{y^*} = 0 \end{cases}$$

$\Longleftrightarrow \begin{pmatrix} \bar{\phi} \\ \bar{\psi} \end{pmatrix}$ is completely canonical, by Proposition 7.1. Hence the assertion.

7.3. <u>The transformation</u> $\begin{pmatrix} \bar{\phi} \\ \bar{\psi} \end{pmatrix}$ <u>is completely canonical</u>

$\Longleftrightarrow$ $x^*dy - \bar{x}^*d\bar{y}$ <u>is the exact differential of some real-valued function of</u> (x,y)

$\Longleftrightarrow$ $y^*dx + \bar{x}^*d\bar{y}$ <u>is the exact differential of some real-valued function of</u> (x,y)

$\Longleftrightarrow$ $y^*dx - \bar{y}^*d\bar{x}$ <u>is the exact differential of some real-valued function of</u> (x,y)

$\Longleftrightarrow$ $x^*dy + \bar{y}^*d\bar{x}$ <u>is the exact differential of some real-valued function of</u> (x,y).

Proof. Add to the differential form $x^* dy - \bar{x}^* d\bar{y}$, in turn, the exact differentials

$$-d(x^* y),$$
$$-d(x^* y) + d(\bar{y}^* \bar{x}),$$
$$d(\bar{y}^* \bar{x}).$$

7.4. Lemma. Let the transformation of local coordinates $z \to \bar{z}$ transform the differential from $Z^*(z)\, dz$ into the differential from $\bar{Z}^*(\bar{z})d\bar{z}$. Then,

$$Z^* dz \text{ is an exact differential}$$
$$\iff \bar{Z}^* d\bar{z} \text{ is an exact differential.}$$

Proof. A straight-forward verification shows that Z_{z^*} is symmetric $\iff$ $\bar{Z}_{\bar{z}^*}$ is symmetric.

7.5 Let

$$\begin{cases} \bar{x} = \bar{\phi}(x,y), \\ \bar{y} = \bar{\psi}(x,y) \end{cases}$$

be a transformation of local coordinates with

$$\det \bar{\psi}_{x^*} \neq 0,$$

so that x and $\bar{x}$ may be considered as locally defined functions of the independent variable $(y,\bar{y})$. Then,

$$\begin{pmatrix} \bar{\phi} \\ \bar{\psi} \end{pmatrix} \text{ is completely canonical}$$

$$\iff \begin{pmatrix} \bar{\phi} \\ \bar{\psi} \end{pmatrix} \text{ is implicitly given by}$$

$$\begin{cases} x = \sigma_y\ (y,\bar{y}), \\ -\bar{x} = \sigma_{\bar{y}}\ (y,\bar{y}) \end{cases}$$

<u>where</u> σ <u>is some function of</u> $(y,\bar{y})$ <u>with</u>

$$\det \sigma_{\bar{y}y^*} \neq 0.$$

<u>Proof.</u>

$$\begin{pmatrix} \bar{\phi} \\ \bar{\psi} \end{pmatrix} \text{ is completely canonical}$$

$\Longleftrightarrow$ $x^*dy - \bar{x}^*d\bar{y}$ is the exact differential of some real function of (x,y) (§7.2)

$\Longleftrightarrow$ $x^*dy - \bar{x}^*d\bar{y}$ is the exact differential of some real function of $(y,\bar{y})$ (§7.4)

$$\Longleftrightarrow \begin{cases} x = \sigma_y(y,\bar{y}), \\ -\bar{x} = \sigma_{\bar{y}}(y,\bar{y}), \end{cases} \text{ and } \sigma_{\bar{y}y^*} = x_{\bar{y}^*} = \bar{y}_{x^*}^{-1} = \bar{\psi}_{x^*}^{-1}.$$

<u>Corollary.</u>

$$\begin{cases} \sigma_{\bar{y}y^*} = \bar{\psi}_{x^*}^{-1}, \\ \sigma_{\bar{y}\bar{y}^*} = -\psi_{\bar{x}^*}^{-1}. \end{cases}$$

<u>7.6</u>. By the preceding proposition, any "<u>generating function</u>" σ yields a completely canonical transformation. However, not every completely canonical transformation is obtained from a generating function: exceptions are completely canonical transformations with $\det \bar{\psi}_{x^*} = 0$, or $\det \psi_{\bar{x}^*} = 0$. An example is the transformation

$$\begin{cases} \bar{x} = x + y, \\ \bar{y} = y, \end{cases}$$

from E^2 onto E^2.

<u>7.7. Corollary to Proposition 7.5</u>: To each of the four differential forms given in Proposition 7.3 there corresponds a real function σ generating a completely canonical transformation. The functions and transformations read:

(i) $(y,\bar{y}) \to \sigma(y,\bar{y})$, $\det \sigma_{\bar{y}y^*} \neq 0$:

$$\begin{cases} x = \sigma_y(y,\bar{y}), \\ -\bar{x} = \sigma_{\bar{y}}(y,\bar{y}). \end{cases}$$

(ii) $(x,\bar{y}) \to \sigma(x,\bar{y})$, $\det \sigma_{x\bar{y}^*} \neq 0$:

$$\begin{cases} y = \sigma_x(x,\bar{y}), \\ \bar{x} = \sigma_{\bar{y}}(x,\bar{y}). \end{cases}$$

(iii) $(x,\bar{x}) \to \sigma(x,\bar{x})$, $\det \sigma_{x\bar{x}^*} \neq 0$:

$$\begin{cases} y = \sigma_x(x,\bar{x}), \\ -\bar{y} = \sigma_{\bar{x}}(x,\bar{x}). \end{cases}$$

(iv) $(y,\bar{x}) \to \sigma(y,\bar{x})$, $\det \sigma_{y\bar{x}^*} \neq 0$:

$$\begin{cases} x = \sigma_y(y,\bar{x}), \\ \bar{y} = \sigma_{\bar{x}}(y,\bar{x}). \end{cases}$$

The transformations (i) and (iii) are of essentially the same kind, and so are the transformations (ii) and (iv): thus there are only two kinds of transformations.

To each transformation there corresponds a theorem analogous to Proposition 7.5.

7.8. Proposition 7.7 yields, in particular, linear completely canonical transformations:

Let C be any non-singular (m×m)-matrix of real numbers. Then the linear transformations

$$\text{(i)} \quad \begin{pmatrix} \bar{x} \\ \bar{y} \end{pmatrix} = \begin{pmatrix} 0 & -C^* \\ C^{-1} & 0 \end{pmatrix} \begin{pmatrix} x \\ y \end{pmatrix} ,$$

$$\text{(ii)} \quad \begin{pmatrix} \bar{x} \\ \bar{y} \end{pmatrix} = \begin{pmatrix} C^* & 0 \\ 0 & C^{-1} \end{pmatrix} \begin{pmatrix} x \\ y \end{pmatrix} ,$$

$$\text{(iii)} \quad \begin{pmatrix} \bar{x} \\ \bar{y} \end{pmatrix} = \begin{pmatrix} 0 & C^{-1} \\ -C^* & 0 \end{pmatrix} \begin{pmatrix} x \\ y \end{pmatrix} , \quad \text{and}$$

$$\text{(iv)} \quad \begin{pmatrix} \bar{x} \\ \bar{y} \end{pmatrix} = \begin{pmatrix} C^{-1} & 0 \\ 0 & C^* \end{pmatrix} \begin{pmatrix} x \\ y \end{pmatrix}$$

are completely canonical.

Their generating functions are, in turn,

(i) $y^* C \bar{y}$,

(ii) $x^* C \bar{y}$,

(iii) $x^* C \bar{x}$,

(iv) $y^* C \bar{x}$.

§8. Jacobi's Partial Differential Equation

8.1. The method of constructing time-independent conpletely canonical transformations by means of generating functions can be extended to a method of constructing time-dependent completely canonical transformations.

The general condition for a real function $(y,\bar{y},t) \to \sigma(y,\bar{y},t)$ to be the generating function of a time-dependent completely canonical transformation, however, will not be discussed here; instead, for any given Hamilton function H a real function σ will be determined such that the corresponding transformation $\bar{\chi}$ transforms H into a particularly simple Hamilton function $\bar{K}(\bar{z},t)$, namely the function

$$\bar{K}(\bar{z},t) = 0 \text{ for all admitted } (\bar{z},t).$$

Thus "the dynamical system is transformed to rest"; the general solution of the Hamilton equation

$$\frac{d\bar{z}}{dt} = J\,\bar{K}_{\bar{z}}\,(\bar{z},t)$$

then reads

$$\bar{z} = c$$

where c denotes an arbitrary constant (2m)-dimensional vector belonging to the appropriate domain. The general solution of the original Hamilton equation

$$\frac{dz}{dt} = J\,H_z(z,t),$$

therefore is,

$$z = \chi(c,t),$$

where χ again denotes the local inverse of $\bar{\chi}$ for fixed values of the time variable t.

<u>8.2.</u> <u>Let the time-dependent transformation of local coordinates</u>

$$\begin{cases} \bar{x} = \bar{\phi}(x,y,t) \\ \bar{y} = \bar{\psi}(x,y,t) \end{cases}$$

be implicitly given by

$$\begin{cases} x = \sigma_y(y,\bar{y},t) \\ -\bar{x} = \sigma_{\bar{y}}(y,\bar{y},t) \end{cases}$$

where σ is a real function of $(y,\bar{y},t)$ with

$$\det \sigma_{y\,\bar{y}*} \neq 0.$$

(Thus x and $\bar{x}$ may be considered as locally defined functions of $(y,\bar{y},t)$.) Further, let H be a given local Hamilton function depending on (x,y,t). Then, $\bar{\chi} \underset{\text{Def.}}{=} \begin{pmatrix} \bar{\phi} \\ \bar{\psi} \end{pmatrix}$ transforms the Hamilton equation $dz/dt = JH_z(z,t)$ into the Hamilton equation $d\bar{z}/dt = 0 \Longleftrightarrow \sigma$ is a "complete solution" of "Jacobi's partial differential equation"

$$\sigma_t - H(\sigma_y,y,t) = 0,$$

i.e., a solution depending on an m-vector parameter $\bar{y}$ such that the above condition $\det \sigma_{y\bar{y}*} \neq 0$ is satisfied.

Proof.

(1) $\bar{\chi}$ transforms the equation $dz/dt = J\,H_z$ into the equation $d\bar{z}/dt = 0$

$$\Longleftrightarrow J^*\,\bar{\chi}_t(\chi,t) = -\bar{H}_{\bar{z}}(\bar{z},t) \qquad (\S 6.4\ )$$

$$\Longleftrightarrow J\,\bar{\chi}_t(z,t) = \bar{\chi}_{z*}{}^* H_z$$

$$\Longleftrightarrow \bar{\chi}_{z*}{}^* J\,\bar{\chi}_t = H_z$$

$$\Longleftrightarrow \begin{cases} \bar{\phi}_{x*}{}^*\,\bar{\psi}_t - \bar{\psi}_{x*}{}^*\,\bar{\phi}_t = H_x\,, \\ \bar{\phi}_{y*}{}^*\,\bar{\psi}_t - \bar{\psi}_{y*}{}^*\,\bar{\phi}_t = H_y\,. \end{cases}$$

(2) Differentiate the identities

$$\left.\begin{aligned} x &= \sigma_y(y,\bar{\psi},t) \ , \\ -\bar{\phi} &= \sigma_{\bar{y}}(y,\bar{\psi},t) \ , \end{aligned}\right\} \quad \text{identically in } (x,y,t),$$

with respect to t, x and y; thus,

$$\begin{cases} \bar{\psi}_t = -\sigma_{y\bar{y}*}{}^{-1}\sigma_{yt} \ , \\ \bar{\phi}_t = \sigma_{\bar{y}\bar{y}*}\,\sigma_{y\bar{y}*}{}^{-1}\sigma_{yt} - \sigma_{\bar{y}t} \ , \end{cases}$$

$$\begin{cases} \bar{\psi}_{x*} = \sigma_{y\bar{y}*}{}^{-1} \ , \\ \bar{\phi}_{x*} = -\sigma_{\bar{y}\bar{y}*}\,\sigma_{y\bar{y}*}{}^{-1} \ , \end{cases}$$

$$\begin{cases} \bar{\psi}_{y*} = -\sigma_{y\bar{y}*}{}^{-1}\sigma_{yy*} \ , \\ \bar{\phi}_{y*} = -\sigma_{\bar{y}y*} + \sigma_{\bar{y}\bar{y}*}\,\sigma_{y\bar{y}*}{}^{-1}\sigma_{yy*} \ . \end{cases}$$

(3) Substitute, from these equations, for the derivatives of $\bar{\phi}$ and $\bar{\psi}$ in (1); thus $\bar{\chi}$ transforms $dz/dt = J\,H_z$ into $d\bar{z}/dt = 0$

$$\Longleftrightarrow \begin{cases} \bar{\phi}_{x*}{}^*\,\bar{\psi}_t - \bar{\psi}_{x*}{}^*\,\bar{\phi}_t = H_x \ , \\ \bar{\phi}_{y*}{}^*\,\bar{\psi}_t - \bar{\psi}_{y*}{}^*\,\bar{\phi}_t = H_y \end{cases} \qquad (1)$$

$$\Longleftrightarrow \begin{cases} (\sigma_{y\bar{y}*}{}^*)^{-1}\sigma_{\bar{y}t} = H_x(\sigma_y,y,t) \ , \\ \sigma_{yt} - \sigma_{yy*}(\sigma_{y\bar{y}*}{}^*)^{-1}\sigma_{\bar{y}t} = H_y(\sigma_y,y,t) \ , \end{cases}$$

identically in $(y,\bar{y},t)$, (2)

$$\Longleftrightarrow \begin{cases} (\sigma_t)_{\bar{y}} = (H(\sigma_y,y,t))_{\bar{y}} \ , \\ (\sigma_t)_y = (H(\sigma_y,y,t))_y \end{cases}$$

$\Longleftrightarrow$ $\sigma_t(y,y,t) - H(\sigma_y,y,t)$

is a function α (say) only of t.

For the direction $\Longleftarrow$ in the proof, choose $\alpha = 0$; for the opposite direction, $\Longrightarrow$, replace H by $H + \alpha$. This completes the proof of the proposition.

8.3. Corollary to Proposition 8.2:

The general solution of the Hamilton equation

$$\begin{cases} \dfrac{dx}{dt} = H_y(x,y,t) \ , \\ \dfrac{dy}{dt} = -H_x(x,y,t) \end{cases}$$

reads

$$\begin{cases} x = \phi(a,b,t) \ , \\ y = \psi(a,b,t) \ , \end{cases}$$

where a and b are arbitrary fixed m-vectors (belonging to the appropriate domains.)

Thus the problem of determining the general solution of Hamilton's equation is reduced to that of determining a particular complete solution of Jacobi's equation. Complete solutions can sometimes be found by the method of "separation of variables".

8.4. Jacobi's equation can be applied to the problem of perturbations (§6.7):

Let $(y,\bar{y},t) \to \sigma(y,\bar{y},t)$ be a complete solution of the Jacobi equation

$$\sigma_t - H(\sigma_y,y,t) = 0 \ ,$$

with

$$\det \sigma_{y\bar{y}^*} \neq 0.$$

Then the transformation $(x,y,t) \to (\bar{x},\bar{y},t)$ defined by means of the function σ transforms the Hamilton equation

$$\frac{dz}{dt} = J \cdot (H + P)_z \qquad (\S 6.7)$$

into the Hamilton equation

$$\frac{d\bar{z}}{dt} = J \cdot \bar{P}_{\bar{z}}$$

where

$$\bar{P}(\bar{z},t) \underset{\text{Def.}}{=} P(\phi(\bar{z},t), \psi(\bar{z},t),t).$$

This procedure may be repeated if for P the same conditions hold as for H, etc.

8.5. A "time-independent Jacobi equation" for time-independent Hamilton equations is obtained in the following way. Suppose the function σ is of the form

$$\sigma(y,\bar{y},t) = \bar{b} * \bar{y} \cdot t + \tau(y,\bar{y})$$

where b is some fixed m-vector. Then Propositions 8.2-4 yield analogous propositions involving the function τ. In particular, the time-independent analogue to Jacobi's equation for σ reads:

$$\bar{b} * \bar{y} - H(\tau_y, y) = 0.$$

Usually for $\bar{b}$ a vector is taken one component of which is 1 whereas all the others vanish.

The corresponding transformation of local coordinates $(x,y,t) \to (\bar{x},\bar{y},t)$ is implicitly defined by:

$$\begin{cases} x = \tau_y(y,\bar{y}), \\ -\bar{x} = \bar{b}\, t + \tau_{\bar{y}}(y,\bar{y}). \end{cases}$$

8.6. By §7.5, the function $\tau(y,\bar{y})$ just introduced may also be taken as the generating function of a time-independent transformation of coordinates $(x,y) \to (\bar{x},\bar{y})$, implicitly defined by

$$\begin{cases} x = \tau_y(y,\bar{y}), \\ -\bar{x} = \tau_{\bar{y}}(y,\bar{y}). \end{cases}$$

This transformation transforms the Hamilton function $H(x,y)$ into the Hamilton function

$$\bar{K}(\bar{x},\bar{y}) = \bar{H}(\bar{x},\bar{y}) = \bar{b} * \bar{y} \qquad \text{(identically in } \bar{x},\bar{y}\text{)}.$$

The corresponding Hamilton equations

$$\begin{cases} \dfrac{d\bar{x}}{dt} = \bar{b}\,, \\ \dfrac{d\bar{y}}{dt} = 0\,, \end{cases}$$

have the general solution

$$\begin{cases} \bar{x} = \overset{\circ}{\bar{x}} + \bar{b}\,t\,, \\ \bar{y} = \overset{\circ}{\bar{y}}\,, \end{cases}$$

where $\overset{\circ}{\bar{x}}$ and $\overset{\circ}{\bar{y}}$ denote arbitrary fixed m-vectors. The general solution of the Hamilton equations

$$\begin{cases} dx/dt = H_y, \\ dy/dt = -H_x \end{cases}$$

is obtained by the transformation $(\bar{x},\bar{y}) \to (x,y)$.

For the generating functions of the forms $\sigma(x,\bar{y})$, $\sigma(x,\bar{x})$ and $\sigma(y,x)$, mentioned in §7.7, there are similar results.

References and Notes

§8.1. For time-independent generating functions and their connection with linear differential forms, see Wintner (1952), §§43-46.

§8.2. The term "complete integral" is more usual, though less recommendable, than the term "complete solution" used in the text.

§8.4. For the method of separation of variables (§8.4) see, for example, Courant-Hilbert (1962).

CHAPTER III

HAMILTON'S PRINCIPLE AND FIRST INTEGRALS

§9. Hamilton's Principle, Euler's Equations

9.1. In this section, local Lagrange curves will be given a new interpretation.

Let $t \to x(t)$ be a local Lagrange curve, i.e., a local solution of the Lagrange equations

$$\begin{cases} \frac{d}{dt} L_{\dot{x}} - L_x = 0 , \\ \frac{dx}{dt} = \dot{x} , \end{cases}$$

where L denotes some given local Lagrange function, defined for $x \in G$ (a region in E^m), $\dot{x} \in E^m$, and $t \in T'$. Further, let

$$\left.\begin{aligned} \overset{\circ}{x} &= x(\overset{\circ}{t}), \\ x' &= x(t') \end{aligned}\right\} \qquad \text{(Def.)}$$

where $\overset{\circ}{t}$ and $t' > \overset{\circ}{t}$ are any two fixed instants in the domain of the Lagrange curve, ε be a real parameter defined in some neighbourhood of 0, and $\xi: [\overset{\circ}{t}, t'] \to E^m$ be an arc which starts and ends at 0:

$$\begin{cases} \xi\,(\overset{\circ}{t}) = 0, \\ \xi\,(t') = 0. \end{cases}$$

Then, for all ε sufficiently close to 0, the arc $x + \varepsilon\xi$, connecting the points $\overset{\circ}{x}$ and x', lies entirely in G.

For these values of ε,

$$\int_{\overset{\circ}{t}}^{t'} L(x(t) + \varepsilon\, \xi(t), \frac{dx(t)}{dt} + \varepsilon \frac{d\xi(t)}{dt}, t)\, dt$$

$$= \int_{\underset{\sim}{t}}^{t'} L(x(t), \frac{dx(t)}{dt}, t)dt$$

$$+ \varepsilon. \left[\frac{\partial}{\partial \varepsilon} \int_{\underset{\sim}{t}}^{t'} L(x(t) + \varepsilon \cdot \xi(t), \frac{dx(t)}{dt} + \varepsilon \cdot \frac{d\xi(t)}{dt}, t)dt \right]_{\varepsilon = 0}$$

$$+ 0(\varepsilon^2) \text{ as } \varepsilon \to 0.$$

The coefficient of ε on the right-hand side is called the (first) variation of the functional $\int_{\underset{\sim}{t}}^{t'} L(.,.,t)dt$ at the argument (function) $t \to x(t)$. If this variation vanishes identically, the functional is said to take a stationary value. In particular, if there are relative maxima and minima of the functional, they are stationary values.

9.2 With the preceding definitions the following proposition holds:

The functional $\int_{\underset{\sim}{t}}^{t'} L(.,.,t)dt$ takes a stationary value at the argument x if, and only if, its variation at the argument x vanishes for all admissible functions ξ. This, in turn, vanishes identically if, and only if, the Lagrange derivative of x vanishes identically, i.e., if the function satisfies Lagrange's equations. ("Hamilton's principle").

Lagrange's equations, derived as the condition for stationary values of the functional considered, are also called Euler's equations. The Hamilton-Jacobi theory for Lagrange equations of dynamical systems, expounded in Chapter II, applies to all Euler equations.

Proof.

Compute the variation of the functional $\int_{\underset{\sim}{t}}^{t'} L(x, \frac{dx}{dt}, t)dt$, for an arbitrary given admissible arc $t \to x(t)$. Differentiation with respect to ε at $\varepsilon = 0$ yields

$$\int_{\overset{\circ}{t}}^{t'} \left[L_{x^*}(x(t), \frac{dx(t)}{dt}, t).\ \xi(t) + L_{\dot{x}^*}(x(t), \frac{dx(t)}{dt}, t).\frac{d\xi(t)}{dt} \right] dt.$$

A partial integration of the second term shows that the variation is $\int_{\overset{\circ}{t}}^{t'} -(\Lambda_L x)^*(t).\ \xi(t)dt$ where Λ_L denotes the Lagrange derivative with respect to the given Lagrange function L. If the function x satisfies Euler's equations, the variation of the function vanishes at x, and the functional takes a stationary value.

Conversely, if the variation at x vanishes identically for all admissible functions ξ, choose ξ in such a way that $(\Lambda_L x)^*.\ \xi$ does not change its sign in $[\overset{\circ}{t},t']$. Choose, for instance, $\xi = \Lambda_L x$. The relations $(\Lambda_L x)^*(t)).\ \xi(t) \geq 0$ (for example) for all $t \in [\overset{\circ}{t},t']$ and

$$\int_{\overset{\circ}{t}}^{t'} [(\Lambda_L x)^*(t)].\ \xi(t)dt = 0$$

imply that

$$(\Lambda_L x)(t) = 0$$

for all $t \in [\overset{\circ}{t},t']$, as has been asserted.

References

Courant-Hilbert (1953), Bliss (1962), Gelfand-Fomin (1963).

§10. First Integrals

10.1. Let L be a local Lagrange function which does not depend on the time variable t. Consequently the corresponding Hamilton function H is also independent of the variable t. Further, let $t \to z(t)$ be any local solution of the Hamilton equation

$$\frac{dz}{dt} = JH_z(z).$$

Then,

$$\frac{d}{dt} H(z(t)) = H_{z*}(z(t)) \frac{dz(t)}{dt}$$

$$= H_{z*}(z(t))JH_z(z(t)).$$

The transpose of this equation reads

$$\frac{d}{dt} H(z(t)) = H_{z*}(z(t))J*H_z(z(t))$$

$$= -H_{z*}(z(t))JH_z(z(t))$$

$$= -\frac{d}{dt} H(z(t)).$$

Therefore,

$$\frac{d}{dt} H(z(t)) = 0,$$

$$H(z(t)) = H(z(\overset{\circ}{t})) ;$$

that is, the function H "is constant along the curve z(t)", it is a "(first) integral of the Hamilton equation".

10.2. This observation motivates the following definitions:

Let

$$\frac{dy}{dt} = g(y,t), \quad \begin{cases} y \in G \subseteq E^n , \\ t \in T' , \end{cases}$$

be a given differential equation (cf. §2.6), and $(\overset{\circ}{y},\overset{\circ}{t},t) \to \psi(\overset{\circ}{y},\overset{\circ}{t},t)$ be its general solution satisfying the initial condition

$$\psi(\overset{\circ}{y},\overset{\circ}{t},\overset{\circ}{t}) = \overset{\circ}{y} \quad \text{for all admitted } \overset{\circ}{y},\overset{\circ}{t}, \text{ and } t.$$

Then any non-constant real function F, defined for all admitted (y,t), is called a <u>global (first) integral</u> of the differential equation if

$$F(\psi(\overset{\circ}{y},\overset{\circ}{t},t),t) = F(\overset{\circ}{y},\overset{\circ}{t}) \text{ for all admissible } \overset{\circ}{y},\overset{\circ}{t}, \text{ and } t.$$

If this equation is satisfied for all $(\overset{\circ}{y},\overset{\circ}{t},t)$ belonging to a sufficiently small neighbourhood in $G \times T' \times T'$, then the function F is said to be a *local (first) integral* of the equation.

Evidently every global integral is a local integral, but the converse need not be true. By §10.1, *any time-independent Hamilton function is a global integral of the corresponding Hamilton equation*.

In the following, the term "(first) integral" will be used when the assertions made are true for both kinds of integrals after the necessary modifications concerning the domains of definition have been made. Such modifications will not always be mentioned explicitly.

10.3. We shall now try to survey all the integrals of the differential equation

$$\frac{dy}{dt} = g(y,t), \quad y \in G, \quad t \in T'.$$

First, there is the following immediate consequence from the definition of the notion of a first integral:

Let $F^{(1)}, F^{(2)},\ldots,F^{(k)}$ *be k functionally independent first integrals of this differential equation, and* ϕ *be any non-constant function of* k *scalar variables such that the function* $\phi(F^{(1)},\ldots,F^{(k)})$ *is defined and non-constant in an open subset of* $G \times T'$. *Then the function* $\phi(F^{(1)},\ldots,F^{(k)})$ *is a first integral of the equation*.

10.4. *Let* $\psi(\mathring{y},\mathring{t},t)$ *be the general solution of the differential equation*

$$\frac{dy}{dt} = g(y,t) \ , \begin{cases} y \in G \subseteq E^n , \\ t \in T' , \end{cases}$$

such that

$$\psi(\mathring{y},\mathring{t},\mathring{t}) = \mathring{y} \quad \text{for all} \quad \mathring{y} \in G, \ \mathring{t} \in T' .$$

Suppose that no component of ψ *is constant*. *Then the components of the*

Then,

vector function $\psi(x,t,\overset{0}{t})$, *with an arbitrary, but fixed, value of* $\overset{0}{t}$ *in* T', *are first integrals of the differential equation, and any of its integrals is a function of* $\psi(x,t,\overset{0}{t})$.

Thus there are exactly n functionally independent first integrals.

The proof will be accomplished by means of two lemmas, which are of importance in themselves.

10.5. *Let, with the notation and assumption of* §10.4,

$$\overset{1}{y} = \psi(\overset{0}{y},\overset{0}{t},\overset{1}{t}).$$

Then,

$$\psi(\overset{1}{y},\overset{1}{t},t) = \psi(\overset{0}{y},\overset{0}{t},t)$$

for arbitrary $\overset{0}{y},\overset{0}{t},\overset{1}{t},t$.

Proof. Consider first the special case that the equation is *autonomous*, that is, its right-hand side does not depend on the time-variable t. Thus we write:

$$\frac{dy}{dt} = g(y) \ .$$

Let $\psi(\overset{0}{y},t)$ be the general solution with

$$\psi(\overset{0}{y},0) = \overset{0}{y} \quad \text{for all admitted } \overset{0}{y},$$

and

$$\overset{1}{y} = \psi(\overset{0}{y},\overset{1}{t} - \overset{0}{t})$$

for arbitrary, but fixed, $\overset{0}{y}$ and $\overset{0}{t}$. Then both the functions $\psi(\overset{1}{y},t - \overset{1}{t})$ and $\psi(\overset{0}{y},t - \overset{0}{t})$ satisfy the differential equation, and both take the value $\overset{1}{y}$ at $t = \overset{1}{t}$; hence, by §2.6, they are equal:

$$\psi(\overset{1}{y},t - \overset{1}{t}) = \psi(\overset{0}{y},t - \overset{0}{t})$$

for arbitrary $\overset{\circ}{y}$, $\overset{\circ}{t}$, t. Thus the assertion is proven for autonomous differential equations.

To prove the assertion also in the general case, we establish an autonomous differential equation which, in a sense, is equivalent to the given non-autonomous equation

$$\frac{dy}{dt} = g(y,t)$$

with the general solution $\psi(\overset{\circ}{y},\overset{\circ}{t},t)$: take y and t as functions of a new independent variable s, defined by the autonomous system of differential equations

$$\begin{cases} \dfrac{dy}{ds} = g(y,t), \\ \dfrac{dt}{ds} = 1 \end{cases}$$

with the general solution

$$\begin{cases} y = \psi(\overset{\circ}{y},\overset{\circ}{t},s - \overset{\circ}{s} + \overset{\circ}{t}), \\ t = s - \overset{\circ}{s} + \overset{\circ}{t}, \end{cases}$$

satisfying the initial condition

$$\left.\begin{matrix} y = \overset{\circ}{y}, \\ t = \overset{\circ}{t} \end{matrix}\right\} \text{ if } s = \overset{\circ}{s}.$$

Let

$$\begin{cases} \overset{1}{y} = \psi(\overset{\circ}{y},\overset{\circ}{t},\overset{1}{s} - \overset{\circ}{s} + \overset{\circ}{t}), \\ \overset{1}{t} = \overset{1}{s} - \overset{\circ}{s} + \overset{\circ}{t} \end{cases}$$

for any admissible fixed value of $\overset{1}{s}$. Hence, by the first part of the proof,

$$\begin{cases} \psi(\overset{1}{y},\overset{1}{t},s - \overset{1}{s} + \overset{1}{t}) = \psi(\overset{\circ}{y},\overset{\circ}{t},s - \overset{\circ}{s} + \overset{\circ}{t}), \\ s - \overset{1}{s} + \overset{1}{t} \qquad = \qquad s - \overset{\circ}{s} + \overset{\circ}{t}. \end{cases}$$

From

$$t = s - \mathring{s} + \mathring{t} ,$$

it follows that

$$\psi(\overset{1}{y},\overset{1}{t},t) = \psi(\mathring{y},\mathring{t},t).$$

10.6. *Let* $y = \psi(\mathring{y},\mathring{t},t)$, *with*

$$\psi(\mathring{y},\mathring{t},\mathring{t}) = \mathring{y} , \qquad \begin{cases} \mathring{y} \varepsilon G , \\ \mathring{t} \varepsilon T' , \end{cases}$$

be the general solution of the differential equation

$$\frac{dy}{dt} = g(y,t) , \qquad \begin{cases} y \varepsilon G , \\ t \varepsilon T' . \end{cases}$$

Then the local mapping $\mathring{y} \to y$, *defined by* $y = \psi(\mathring{y},\mathring{t},t)$ *for any fixed values of* $\mathring{t}$ *and* t, *has the unique local inverse*

$$\mathring{y} = \psi(y,t,\mathring{t}).$$

Proof. Substitute, in the equations of Proposition 10.5, t for $\overset{1}{t}$, y for $\overset{1}{y}$, and $\mathring{t}$ for t: hence,

$$y = \psi(\mathring{y},\mathring{t},t)$$

implies

$$\begin{aligned} \psi(y,t,\mathring{t}) &= \psi(\mathring{y},\mathring{t},\mathring{t}) \\ &= \mathring{y} , \end{aligned}$$

as has been asserted.

10.7. The first part of Proposition 10.4 is an immediate consequence of Proposition 10.6. To demonstrate the second part, let $F(x,t)$ be any first integral of the equation

$$\frac{dy}{dt} = g(y,t).$$

Then,

$$F(\psi(\mathring{y},\mathring{t},t),t) = F(\mathring{y},\mathring{t})$$

by the definition of "first integral" (§10.2),

$$F(y,t) = F(\psi(y,t,\mathring{t}),\mathring{t}) \quad \text{by Proposition 10.6,}$$

finally,

$$F(y,t) = \mathring{F}(\psi(y,t,\mathring{t})),$$

where

$$\mathring{F}(y) \underset{\text{Def.}}{=} F(y,\mathring{t}),$$

and this is the second assertion of Proposition 10.4.

10.8. First integrals are characterized by the following proposition: *A non-constant real-valued function* $F(y,t)$, *defined in an appropriate domain, is a first integral of the differential equation*

$$\frac{dy}{dt} = g(y,t)$$

if, and only if, it satisfies "Liouville's equation", i.e., the partial differential equation

$$F_t + g^*(y,t).\ F_y = 0.$$

The proof is similar to that of the special case discussed in §10.1: Consider the derivative

$$\frac{d}{dt}F(\psi(\mathring{y},\mathring{t},t),t)$$

where $\psi(\mathring{y},\mathring{t},t)$ denotes the general solution of the differential equation. This derivative vanishes identically if, and only if, F is a first integral.

Proposition 10.4 implies that any (n+1) first integrals (defined in the sam domain) are functionally dependent. This result is also obvious from Liouville's equation: each of the first integrals considered satisfies Liouville's equation, and hence it follows that the functional determinant of the (n+1) first integrals vanishes identically.

10.9. We now turn to applications of first integrals.

Let $F(y,t)$ be a first integral of the differential equation

$$\frac{dy}{dt} = g(y,t),$$

and suppose that at the point $(\overset{\circ}{y},\overset{\circ}{t}) \varepsilon E^{n+1}$ at least one of the partial derivatives of F does not vanish. Let, for instance,

$$\frac{\partial F}{\partial y_n}(\overset{\circ}{y},\overset{\circ}{t}) \neq 0.$$

Then the equation

$$F(y,t) = \beta$$

where

$$\beta \underset{\text{Def.}}{=} F(\overset{\circ}{y},\overset{\circ}{t})$$

can locally be solved with respect to y_n. The solution is a function of $(y_1,\cdots,y_{n-1})$, t and β. Substitute it, in the differential equation, for y_n. Then the first (n-1) equations of the corresponding system of n scalar differential equations do not contain y_n at all. Thus, by means of the first integral, a new system of (n-1) scalar differential equations has been derived from the original system of n equations. After the new system has been solved, y_n can be determined from the equation $F(y,t) = \beta$.

Thus, a first integral can be used for a local reduction of the differential equation to one of a lower order. In addition, global first integrals may be useful for the qualitative description of the solutions of the equation. They are indispensable in the applications of probability theory to dynamics.

10.10. An autonomous system of n differential equations

$$\frac{dy_i}{dt} = g_i(y) \; , \quad i=1,2,\cdots,n,$$

can, in general, locally be reduced to a system of (n-1) equations.

Suppose, for example, that locally

$$g_n(y) \neq 0;$$

then introduce y_n as a new independent variable by a division:

$$\frac{dy_i}{dy_n} = \frac{g_i(y)}{g_n(y)} \; , \quad i=1,2,\cdots,n-1.$$

After $y_1, \cdots y_{n-1}$ have been determined from these equations as functions of y_n, the differential equation for y_n yields

$$t = \int \frac{dy_n}{g_n(y)} \; .$$

10.11. Any time-independent local first integral of the system

$$\frac{dy_i}{dt} = g_i(y) \; , \quad i=1,\cdots,n \; ,$$

is a local first integral of the system

$$\frac{dy_i}{dy_n} = \frac{g_i(y)}{g_n(y)} \; , \quad i=1,\cdots,n-1 \; ,$$

and conversely.

This is an immediate consequence of the definition of the notion of a first integral, according to which the value of such an integral is "constant along any local solution curve" of the system of differential equations. The assertion can also be read from the corresponding Liouville equation (§10.8).

10.12. It follows that a system of n autonomous differential equations for which a time-independent first integral is known can be reduced to a system of (n-2) equations: first eliminate one of the unknown functions by means of the integral (as in §10.9), and then eliminate, by a division, the time variable (as in §10.10). In particular:

An autonomous Hamiltonian system of 2m differential equations can locally be reduced to a system of (2m-2) equations. Moreover, there is a reduction such that the reduced system is also Hamiltonian.

Proof of the last assertion: Let H(x,y) be the given Hamilton function, and suppose that, for example, its partial derivative with respect to the component y_m of the vector y does not vanish in some open neighbourhood of a given point $(\overset{\circ}{x},\overset{\circ}{y}) \in E^{2m}$. Write "$(\tilde{y},y_m)$" for "y" and "$(\tilde{x},x^m)$" for "x". Thus,

$$\begin{cases} \dfrac{d\tilde{x}}{dt} = H_{\tilde{y}} , & \dfrac{d\tilde{y}}{dt} = -H_{\tilde{x}} , \\[2ex] \dfrac{dx^m}{dt} = H_{y_m} , & \dfrac{dy_m}{dt} = -H_{x^m} , \end{cases}$$

$$(*) \qquad \begin{cases} \dfrac{d\tilde{x}}{dx^m} = \dfrac{H_{\tilde{y}}}{H_{y_m}} , & \dfrac{d\tilde{y}}{dx^m} = - \dfrac{H_{\tilde{x}}}{H_{y_m}} , \\[2ex] & \dfrac{dy_m}{dx^m} = - \dfrac{H_{x^m}}{H_{y_m}} . \end{cases}$$

Now let

$$\beta \underset{\text{Def.}}{=} H(\overset{\circ}{x},\overset{\circ}{y})$$

(assuming that $(\overset{\circ}{x},\overset{\circ}{y})$ is the starting point for the solution curve considered), and solve the equation

$$H(\tilde{x},x^m,\tilde{y},y_m) = \beta$$

locally with respect to y_m:

$$y_m = h(\tilde{x}, x^m, \tilde{y} | \beta).$$

Hence,

$$H(\tilde{x}, x^m, \tilde{y}, h(\tilde{x}, x^m, \tilde{y} | \beta)) = \beta$$

identically in $\tilde{x}, x^m, \tilde{y}$ and β. Differentiate this identity with respect to $\tilde{x}$ and $\tilde{y}$, solve the resulting equations for $h_{\tilde{x}}$ and $h_{\tilde{y}}$, and substitute in (*): thus,

$$\frac{d\tilde{x}}{dx^m} = -h_{\tilde{y}} \, , \quad \frac{d\tilde{y}}{dx^m} = h_{\tilde{x}} \, ,$$

which implies the assertion.

Corollary. The Hamilton function of the reduced system is $-h(\tilde{x}, x^m, \tilde{y} | \beta)$.

10.13. There is the following converse to the last proposition: To any non-autonomous Hamiltonian system of 2m differential equations

(*) $$\frac{dx}{dt} = H_y(x,y,t) \, , \quad \frac{dy}{dt} = -H_x(x,y,t)$$

there corresponds an "extended" Hamiltonian system of 2(m+1) autonomous equations:

(**) $$\begin{cases} \dfrac{dx}{d\tau} = G_y(x,y,s,t), & \dfrac{dy}{d\tau} = -G_x(x,y,s,t), \\ \dfrac{ds}{d\tau} = G_t(x,y,s,t), & \dfrac{dt}{d\tau} = -G_s(x,y,s,t), \end{cases}$$

where

$$G(x,y,s,t) \underset{\text{Def.}}{=} H(x,y,t)-s$$

and τ is a new time variable (which is not essentially different from the original time variable t). The system (*) is obtained from the system (**) by a reduction by means of the first integral G.

For the proof apply the method of §10.12 to the system (**).

10.14. First integrals are also used in Poisson's method of perturbations:

Let

$$\bar{z} = \bar{\chi}(z,t),$$

with

$$\bar{\chi}(z,0) = z \quad \text{for all admitted } z,$$

be the general solution of the differential equation

$$\frac{d\bar{z}}{dt} = \bar{f}(\bar{z},t),$$

and

$$z = \chi(\bar{z},t)$$

be, for any fixed value of t, the local inverse of $\bar{\chi}$. Then the mapping χ transforms the differential equation

$$\frac{d\bar{z}}{dt} = \bar{f}(\bar{z},t) + \bar{g}(\bar{z},t)$$

into the differential equation

$$\frac{dz}{dt} = \chi_{\bar{z}*}(\bar{\chi}(z,t),t)\cdot g(z,t)$$

where

$$g(z,t) \underset{\text{Def.}}{=} \bar{g}(\bar{\chi}(z,t),t).$$

Proof. $\frac{dz}{dt} = \chi_t(\bar{z},t) + \chi_{\bar{z}*}(\bar{z},t)\ \frac{d\bar{z}}{dt}$

$$= \chi_t + \chi_{\bar{z}*}(\bar{f} + \bar{g})$$

$$= (\chi_t + \chi_{\bar{z}*}\bar{f}) + \chi_{\bar{z}*}\bar{g}.$$

The bracket on the right-hand side vanishes, by Proposition 10.8, since χ is, by Proposition 10.4, a first integral of the equation $d\bar{z}/dt = \bar{f}$. Hence the assertion.

<u>10.15</u>. An application of Poisson's method to Hamilton equation yields Proposition 6.7, which thus is seen to be a special case of Proposition 10.14:

Let

$$\bar{f}(\bar{z},t) = J\bar{H}_{\bar{z}}(\bar{z},t)$$

and

$$\bar{g}(\bar{z},t) = J\bar{P}_{\bar{z}}(\bar{z},t).$$

Then the equation

$$\frac{d\bar{z}}{dt} = J\cdot(\bar{H} + \bar{P})_{\bar{z}}$$

is by the transformation χ transformed into

$$\frac{dz}{dt} = \chi_{z^*} J\bar{P}_{\bar{z}}$$

$$= \chi_{z^*} J\chi^*_{z^*} P_z \ .$$

Since $\bar{\chi}$ is the general solution (with distinguished initial condition) of a Hamilton equation (viz. $d\bar{z}/dt = J\bar{H}_{\bar{z}}$), it is, by Proposition 6.3, completely canonical, and so is its inverse, χ. Therefore,

$$\chi_{z^*} J\chi^*_{z^*} = J,$$

by Proposition 5.6, and

$$\frac{dz}{dt} = J\cdot P_z,$$

as has been asserted in Proposition 6.7.

10.16. Lagrange's method of perturbations, though in derivation and mode of application different from Poisson's method, is essentially identical with it:

Using the notation of §10.14, we have

$$\frac{d\bar{z}}{dt} = \frac{d}{dt}\bar{\chi}(z,t) = \bar{f} + \bar{g},$$

$$\bar{\chi}_t + \bar{\chi}_{z*}\frac{dz}{dt} = \bar{f} + \bar{g}.$$

By the definition of $\bar{\chi}$,

$$\bar{\chi}_t = \bar{f} ;$$

hence,

$$\bar{\chi}_{z*}\frac{dz}{dt} = \bar{g}$$

or

$$\frac{dz}{dt} = \bar{\chi}_{z*}^{-1}(z,t)\cdot g(z,t).$$

But because $\bar{\chi}_{z*}^{-1} = \underline{\chi}_{z*}$, this is equivalent to Poisson's formula of §10.14,

$$\frac{dz}{dt} = \underline{\chi}_{z*}\cdot g.$$

In concrete applications, Poisson's formula may be preferable because there is no need to compute the inverse of a matrix; instead, the inverse of the solution function $\bar{\chi}$ is to be determined.

10.17. The question now arises how first integrals, in particular, global first integrals, can actually be found.

If the Hamilton function H(x,y,t) does not depend on a given component of the vector x, or the vector y, then, as the Hamilton equations show, the corresponding component of the vector y or x, resp., is a first integral of the equations. (Such a component is called an ignorable coordinate of the system. The purpose of the method of §8.6 is to introduce ignorable coordinates.) There is an analogous proposition for Lagrange equations.

10.18. The following proposition makes it sometimes possible to construct, from two given time-independent global integrals F and F' of the Hamilton equation

$$\frac{dz}{dt} = JH_z(z),$$

an additional first integral:

Let F(z) *and* F'(z) *be any two time-independent integrals, and* $t \to z(t)$ *be any solution of the Hamilton equation*

$$\frac{dz}{dt} = JH_z(z).$$

Then the Poisson bracket expression $F_{z*}JF'_z$ *is constant along the solution curve* z(t).

That is,

$$\frac{d}{dt}((F_{z*}JF'_z)(z(t))) = 0 \quad \text{for all admitted values of } t.$$

Thus, if the function $F_{z*}JF'_z$ is non-constant in its domain of definition, it is a first integral of the Hamilton equation. This integral may, or may not, be functionally dependent on the integrals F and F'. In the latter case, an essentially new integral has been found. An example will be mentioned in §11.9.

Proof. The omitted arguments of the functions are always z(t).

(1) $$\frac{d}{dt}(F_{z*}JF'_z) = (\frac{d}{dt}F_{z*})JF'_z + F_{z*}J(\frac{d}{dt}F'_z)$$

$$= -F'_{z*}J\frac{dF_z}{dt} + F_{z*}J\frac{dF'_z}{dt}.$$

(2) $$\frac{dF_z}{dt} = F_{zz*}JH_z .$$

(3) $\frac{d}{dt}(F_{z^*}JF'_z) = -F'_{z^*}JF_{zz^*}JH_z + F_{z^*}JF'_{zz^*}JH_z.$

(4) $(F_{z^*}JH_z)_z = F_{z^*z}JH_z - H_{z^*z}JF_z.$

(5) $F_{z^*}JH_z = 0.$ (Hyp.) (§10.8)

(6) $F_{zz^*}JH_z = H_{zz^*}JF_z.$ (4) (5)

(7) $\frac{d}{dt}(F_{z^*}JF'_z) = -F'_zJH_{zz^*}JF_z + F_{z^*}JH_{zz^*}JF'_z.$ (3) (6)

(8) $F_{z^*}JH_{zz^*}JF'_z = (F_{z^*}JH_{zz^*}JF'_z)^*$

$= F'_{z^*}JH_{z^*z}JF_z$

$= F'_{z^*}JH_{zz^*}JF_z.$

(9) $\frac{d}{dt}(F_{z^*}JF'_z) = 0.$ (7) (8),

as has been asserted.

10.19. For a dynamical system of particles the equations of motion read

$$\mu_i \frac{d^2x}{dt^2} = -\frac{\partial}{\partial x_i} U(x_1,x_2,\ldots,x_p;t), \quad i=1,2,\ldots,p.$$

(§2.6) If U is independent of the time variable t, the Hamilton function

$$\tfrac{1}{2}\sum_{i=1}^{n} \frac{1}{\mu_i} y_i^* y_i + U(x_1,\ldots,x_p)$$

is a first integral of the equations of motion, the "energy integral". If U depends on t, then, in general, no similar integral is known. There is, however, one class of time-dependent potential functions for which there is an analogue to the energy integral.

For constructing this integral it is convenient to use the vectors

$$\bar{x}_i = x_i\sqrt{\mu_i} \quad , \quad i=1,2,\ldots,p \ ,$$

as position vectors of the particles. Introducing then in the equations of motions and dropping the bars yields the equations

$$\frac{d^2x_i}{dt^2} = -\frac{\partial U}{\partial x_i}(x_1,\ldots,x_p) \ , \quad i=1,2,\ldots,p \ ,$$

(with a slight change of the meaning of the symbol "U"). Take $(x_1,\ldots,x_p)$ as a partition of the (p.n)-vector x: thus,

$$\frac{d^2x}{dt^2} = -U_x(x).$$

We now apply the following "space-time transformation" ("Schürer's transformation"):

$$\begin{cases} x = \dfrac{\xi}{\phi(\tau)} \ , \\ \dfrac{d\tau}{dt} = (\phi(\tau))^2 \end{cases}$$

where ϕ denotes an arbitrary positive real function defined on a suitable (non-empty) open interval of the real line. The corresponding transformation between the tangent vectors (velocity vectors) $\dot{x}$ and ξ' reads

$$\dot{x} = \phi(\tau)\xi' - \frac{d\phi(\tau)}{d\tau}\xi \ .$$

It is obtained by applying Schürer's transformation to any function $t \rightarrow x(t)$. A straight-forward verification now confirms the following proposition:

The autonomous system of differential equations

$$\begin{cases} \dfrac{dx}{dt} = \dot{x} \ , \\ \dfrac{d\dot{x}}{dt} = -U_x(x) \end{cases}$$

is, by Schürer's transformation, transformed into the non-autonomous system

$$\begin{cases} \dfrac{d\xi}{d\tau} = \xi' \ , \\ \dfrac{d\xi'}{d\tau} = -\Omega_\xi(\xi,\tau) \end{cases}$$

where

$$\Omega(\xi,\tau) \underset{\text{Def.}}{=} -\tfrac{1}{2}\left[\phi(\tau)\right]^{-1} \frac{d^2\phi(\tau)}{d\tau^2}\ \xi*\xi + \left[\phi(\tau)\right]^{-2} \cdot U\left(\frac{\xi}{\phi(\tau)}\right) .$$

To the energy integral

$$\tfrac{1}{2}\ \dot{x}*\dot{x} + U(x)$$

of the autonomous system there corresponds the first integral

$$\tfrac{1}{2}\left(\phi(\tau)\xi' - \frac{d\phi(\tau)}{d\tau}\ \xi\right)^* \cdot \left(\phi(\tau)\xi' - \frac{d\phi(\tau)}{d\tau}\ \xi\right) + U\left(\frac{\xi}{\phi(\tau)}\right)$$

of the non-autonomous system.

Notes and References

§10.9. For the application of global first integrals in statistical mechanics see, for example, Khinchin (1949), Kurth (1952), (1957b), (1960a); for the qualitative description of the motions of dynamical systems by means of global first integrals see, e.g., Wintner (1952), Siegel (1956), Kurth (1971a). For further applications of local integrals see Whittaker (1961).

§10.14. Applications of Poisson's method are given in Kurth (1959).

§10.17. As to a method of constructing first integrals, cf. Kurth (1971b), (1973b).

§10.19. See Schürer (1943).

§11. Noether's Theorem

11.1. This theorem will, under certain conditons, enable us to derive global first integrals of Euler's equations from their Lagrange function. Our discussion will be local; but it will become apparent that the results obtained are invariant under admissible coordinate transformations and can, therefore, be extended to the whole Lagrange manifold.

Let $t \to \phi(t)$ be any curve in a given admissible system of local coordinates. We define a family of curve transformations by the formulae

$$\begin{cases} \bar{x} = \Phi(x,\dot{x},t,\varepsilon) , \\ \bar{t} = \Phi_0(x,\dot{x},t,\varepsilon) \end{cases}$$

where ε denotes a real parameter defined in a neighbourhood of 0, and Φ and Φ_0 are given functions of ε, t, x, and $\dot{x}$. It is further assumed that

$$\begin{cases} \Phi(x,\dot{x},t,0) = x , \\ \Phi_0(x,\dot{x},t,0) = t . \end{cases}$$

Then, by Taylor's theorem

$$\left.\begin{aligned} \bar{x} &= x + \varepsilon\cdot\xi(x,\dot{x},t) + 0(\varepsilon^2) , \\ \bar{t} &= t + \varepsilon\cdot\tau(x,\dot{x},t) + 0(\varepsilon^2) \end{aligned}\right\} \quad \text{as} \quad \varepsilon \to 0 ,$$

where

$$\xi = (\partial\Phi/\partial\varepsilon)_{\varepsilon=0} , \quad \tau = (\partial\Phi_0/\partial\varepsilon)_{\varepsilon=0} . \qquad \text{(Def.)}$$

Finally it is supposed that, for any admissible set of fixed values of x, $\dot{x}$ and ε, the function $t \to \bar{t}$ is one-to-one.

We now define the transform $\bar{t} \to \bar{\phi}(\bar{t})$ of the curve $t \to \phi(t)$ by

$$\begin{cases} \bar{\Phi}(\bar{t}) = \Phi(\phi(t), \dfrac{d\phi(t)}{dt}, t, \varepsilon) , \\ \bar{t} = \Phi_0(\phi(t), \dfrac{d\phi(t)}{dt}, t, \varepsilon). \end{cases}$$

11.2. In order to have a simple notation, we shall in the following write x instead of ϕ. Confusion between the independent variable x and the function $t \to x(t)$ is avoided by writing x, $\dot{x}$ when x and $\dot{x}$ denote independent variables, and x, $\frac{dx}{dt}$ when x denotes a function of t.

For $\varepsilon \to 0$ the transform of the curve $t \to x(t)$ is described by the asymptotic formulas

$$\begin{cases} \bar{x} = x + \varepsilon \cdot \xi(x, \dfrac{dx}{dt}, t) + O(\varepsilon^2) , \\ \bar{t} = t + \varepsilon \cdot \tau(x, \dfrac{dx}{dt}, t) + O(\varepsilon^2) , \end{cases}$$

and a straight-forward computation shows that

$$\frac{d\bar{x}}{d\bar{t}} = \frac{dx}{dt} + \varepsilon \cdot \left(\frac{d\xi}{dt} - \frac{dx}{dt}\frac{d\tau}{dt} \right) + O(\varepsilon^2) .$$

Here $d\xi/dt$ stands for

$$\frac{\partial \xi}{\partial t}(x, \frac{dx}{dt}, t) + \xi_{x^*}(x, \frac{dx}{dt}, t) \cdot \frac{dx}{dt} + \xi_{\dot{x}^*}(x, \frac{dx}{dt}, t) \cdot \frac{d^2x}{dt^2} ,$$

and there is an analogous expression for $d\tau/dt$.

11.3. Hence, for $\varepsilon \to 0$,

$$L(\bar{x}, \frac{d\bar{x}}{d\bar{t}}, \bar{t}) = L\left(x + \varepsilon \cdot \xi(x, \frac{dx}{dt}, t) + O(\varepsilon^2), \frac{dx}{dt} + \varepsilon \cdot \left(\frac{d\xi}{dt} - \frac{dx}{dt}\frac{d\tau}{dt} \right) + O(\varepsilon^2), \right.$$
$$\left. t + \varepsilon \cdot \tau + O(\varepsilon^2) \right)$$

and, by Taylor's theorem,

$$L(\bar{x}, \frac{d\bar{x}}{d\bar{t}}, \bar{t}) = L(x, \frac{dx}{dt}, t)$$

$$+ \varepsilon\{L_{x^*} \cdot \xi + L_{\dot{x}^*} \cdot (\frac{d\xi}{dt} - \frac{dx}{dt}\frac{d\tau}{dt}) + L_t \cdot \tau\} + 0(\varepsilon^2)$$

where $L_t = \partial L/\partial t$ (Def.)

11.4. Let t_0 and $t_1 > t_0$ be any two real numbers such that the arc $t \to x(t)$, with $t_0 \leq t \leq t_1$, lies in the coordinate neighbourhood considered, and let, for i = 0, 1,

$$\bar{t}_i = t_i + \varepsilon \cdot \tau(x(t_i), \frac{dx}{dt}(t_i), t_i) + 0(\varepsilon^2).$$

Then,

$$\int_{\bar{t}_0}^{\bar{t}_1} L(\bar{x}, \frac{d\bar{x}}{d\bar{t}}, \bar{t})d\bar{t} = \int_{t_0}^{t_1} L(x + \varepsilon \cdot \xi(x, \frac{dx}{dt}, t), \frac{dx}{dt} + \varepsilon \cdot (\frac{d\xi}{dt} - \frac{dx}{dt}\frac{d\tau}{dt}),$$

$$t + \varepsilon \cdot \tau) \frac{d\bar{t}}{dt} dt + 0(\varepsilon^2)$$

$$= \int_{t_0}^{t_1} L(x, \frac{dx}{dt}, t)dt$$

$$+ \varepsilon \cdot \int_{t_0}^{t_1} (L_{x^*} \cdot \xi + L_{\dot{x}^*} (\frac{d\xi}{dt} - \frac{dx}{dt}\frac{d\tau}{dt})$$

$$+ L_t \cdot \tau + L \frac{d\tau}{dt})dt + 0(\varepsilon^2)$$

Hence, by partial integrations,

$$\int_{\bar{t}_0}^{\bar{t}_1} L(\bar{x}, \frac{d\bar{x}}{d\bar{t}}, \bar{t})d\bar{t} - \int_{t_0}^{t_1} L(x, \frac{dx}{dt}, t)dt$$

$$= \varepsilon \cdot \int_{t_0}^{t_1} (L_{x^*} - \frac{d}{dt} L_{\dot{x}^*}) \cdot \xi \quad dt$$

$$+ \varepsilon \cdot \int_{t_0}^{t_1} (L_t + \frac{d}{dt} (L_{\dot{x}^*} \cdot \frac{dx}{dt}) - \frac{d}{dt} L) \cdot \tau \, dt$$

$$+ \varepsilon \cdot \left[L_{\dot{x}^*} \cdot \xi - L_{\dot{x}^*} \frac{dx}{dt} \cdot \tau + L \cdot \tau \right]_{t_0}^{t_1}$$

$$+ O(\varepsilon^2).$$

Carrying out the total differentiations in the second term of the right-hand side, we finally obtain:

$$\int_{\bar{t}_0}^{\bar{t}_1} L(\bar{x}, \frac{d\bar{x}}{d\bar{t}}, \bar{t}) d\bar{t} - \int_{t_0}^{t_1} L(x, \frac{dx}{dt}, t) dt$$

$$= \varepsilon \cdot \int_{t_0}^{t_1} (L_{x^*} - \frac{d}{dt} L_{\dot{x}^*}) \cdot (\xi - \tau \cdot \frac{dx}{dt}) dt$$

$$+ \varepsilon \cdot \left[L_{\dot{x}^*} \cdot \xi + (L - L_{\dot{x}^*} \cdot \frac{dx}{dt}) \cdot \tau \right]_{t_0}^{t_1}$$

$$+ O(\varepsilon^2).$$

<u>11.5</u>. We now define:

The functional $\int_{t_0}^{t_1} L(x, \frac{dx}{dt}, t) dt$ is said to be <u>invariant under the family of curve transformations</u>

$$\begin{cases} \bar{x}(\bar{t}) = \Phi(x(t), \frac{dx}{dt}(t), t, \varepsilon), \\ \bar{t} = \Phi_0(x(t), \frac{dx}{dt}(t), t, \varepsilon) \end{cases}$$

if, for all admissible t_0, t_1 and ε ,

$$\int_{\bar{t}_0}^{\bar{t}_1} L(\bar{x}, \frac{d\bar{x}}{d\bar{t}}, \bar{t})d\bar{t} = \int_{t_0}^{t_1} L(x, \frac{dx}{dt}, t)dt$$

where $\bar{t}_0$ and $\bar{t}_1$ are the values of $\bar{t}$ corresponding to t_0 and t_1 .

It is said to be <u>asymptotically invariant</u> if

$$\int_{\bar{t}_0}^{\bar{t}_1} L(\bar{x}, \frac{d\bar{x}}{d\bar{t}}, \bar{t})d\bar{t} = \int_{t_0}^{t_1} L(x, \frac{dx}{dt}, t)dt + O(\varepsilon^2) \text{ as } \varepsilon \to 0.$$

Obviously invariance implies asymptotic invariance.

The following proposition, called <u>Noether's theorem</u>, is an immediate consequence of these definitions and the result of §11.4:

<u>Let the functional</u> $\int_{t_0}^{t_1} L(x, \frac{dx}{dt}, t)dt$ <u>be asymptotically invariant under the family of curve transformations considered</u>. <u>Then the function</u>

$$L_{\dot{x}^*} \xi + (L - L_{\dot{x}^*} \dot{x}) \cdot \tau ,$$

<u>with arguments</u> $x, \dot{x}, t$, <u>is a first integral of Euler's equations of motion</u>

$$\begin{cases} \dfrac{d}{dt} L_{\dot{x}} = L_x , \\ \\ \dfrac{d}{dt} x = \dot{x} . \end{cases}$$

Note that, by §4.2, this integral can in a more symmetric manner also be written as $L_{\dot{x}^*} \xi - H \cdot \tau$, or $y^*\xi - H \cdot \tau$.

<u>Proof</u>. Introduce the hypothesis of invariance into the final formula

of §11.4, divide by $\varepsilon \neq 0$, and let $\varepsilon \to 0$: thus

$$\int_{t_0}^{t_1} (L_{x*} - \frac{d}{dt} L_{\dot{x}*}) (\xi - \tau \cdot \frac{dx}{dt})dt + \left[L_{\dot{x}*} \cdot \xi + (L - L_{\dot{x}*} \frac{dx}{dt}) \cdot \tau \right]_{t_0}^{t_1} = 0.$$

For any solution $x(t)$ of Euler's equations the first term vanishes. Keep t_0 fixed and take t_1 as a variable: it follows that the function $L_{\dot{x}*} \cdot \xi + (L - L_{\dot{x}*} \cdot \dot{x}) \cdot \tau$ of the independent variables x, $\dot{x}$ and t is constant along any solution of Euler's equations; that is, it is a first integral of these equations - as has been asserted.

11.6. We consider a few examples.

First, suppose that the function L is independent of the time variable t. Then the functional $\int_{t_0}^{t_1} L(x, \frac{dx}{dt})dt$ is invariant under the transformations

$$\begin{cases} \bar{x}(\bar{t}) = x(t) , \\ \bar{t} = t + \varepsilon. \end{cases}$$

Therefore,

$$\xi(x,\dot{x},t) = 0, \quad \tau(x,\dot{x},t) = 1 ,$$

and $L - L_{\dot{x}*} \dot{x}$ is a first integral of Euler's equations. Correspondingly, H is (as we already know from §10.1) a first integral of Hamilton's equations $dx/dt = H_y$, $dy/dt = -H_x$; it is called the "integral of energy."

11.7. Secondly, suppose that the functional $\int_{t_0}^{t_1} L\,d\,t$ is invariant under a family of translations:

$$\bar{x} = x + a\,\varepsilon ,$$

where a denotes some fixed m-vector, and ε is an arbitrary real number.

This is the case for dynamical systems of particles, as considered in §2.4. It follows that the scalar function $(L_{\dot{x}^*} \cdot a)$ is a first integral of Euler's equations.

If the functional is invariant for several linearly independent vectors a, then each of them yields a first integral. Thus for dynamical systems of particles in the n-dimensional Euclidean space E^n there are n such "integrals of momentum" $(L_{\dot{x}^*} \cdot a)$. Hence, in this case, the components of $L_{\dot{x}}$ are first integrals.

11.8. The "integrals of the centre of masses" of particle dynamics are not obtained from Noether's theorem: their existence is too closely related to the particular form of the gradient $L_{\dot{x}}$ for such a system.

We can give a general sufficient condition for the existence of a first integral corresponding to an integral of centre of masses. Suppose that, firstly, the assumption of §11.7 holds so that

$$L_{\dot{x}^*} \cdot a = \sum_{i=1}^{n} \frac{\partial L}{\partial \dot{x}^i} a^i$$

is a first integral of Euler's equation,

and, secondly, $L_{\dot{x}}$ is of the special form

$$L_{\dot{x}}(x,\dot{x}) = \begin{pmatrix} c_1(x^1)\dot{x}^1 \\ c_2(x^2)\dot{x}^2 \\ \vdots \\ c_n(x^n)\dot{x}^n \end{pmatrix}$$

where c_i, for $i=1,2,\cdots,m$, is a given function of x^i.

Define the functions C_i by

$$C_1(x^1) = \int c_1(x^1)dx^1 \text{ , etc.}$$

Then

$$\sum_{i=1}^{n} \{C_i(x^i) - t \cdot \dot{x}^i \cdot c_i(x^i)\} \cdot a^i$$

is a first integral of Euler's equations.

Proof. Substitute for x any solution of Euler's equations, differentiate with respect to t, and take into account that $L_{\dot{x}^*} \cdot a$ is a first integral.

11.9. Finally suppose that a functional $\int_{t_0}^{t_1} L(x, \frac{dx}{dt}, t)dt$ is invariant under a family of rotations

$$\bar{x} = A(\varepsilon) \cdot x$$

where $A(\varepsilon)$ denotes an orthogonal matrix depending on the parameter ε, and

$$A(0) = I_m \text{ (the identity matrix).}$$

Thus

$$A(\varepsilon) = I_m + \varepsilon C + O(\varepsilon^2),$$

and the condition that

$$AA^* = I_m$$

implies that

$$C + C^* = 0 \ ;$$

that is, the matrix C is skew-symmetric. The corresponding first integral reads

$$L_{\dot{x}^*} \cdot Cx \ .$$

For dynamical systems of particles the first "integrals of the moment of momentum" are obtained in this way.

Thus all the first integrals of classical dynamics have been derived.

For dynamical systems of particles in the sense of §2.4, three functionally independent integrals of the moment of momentum exist, and each of them can be derived from both the others by means of Proposition 10.17.

Notes and References

§11.4. A more geometrical derivation of the basic formula is given in Gelford - Fomin (1963).

§12. Stability

12.1. In this section an application and a generalization of the notion of a first integral will be discussed.

Consider, first, a dynamical system whose Hamilton function H is independent of the time variable t, and suppose that the constant 0 is a solution of the Hamilton equation

$$\frac{dz}{dt} = JH_z(z);$$

that is, 0 is an "equilibrium solution" of the Hamilton equations. The question to be discussed is whether or not this solution is "stable".

The term "stable" has several meanings. Here it will be used in the sense of "Liapunov stability": a solution $t \to z_0(t)$ of a given differential equation

$$\frac{dz}{dt} = g(z,t)$$

is said to be (Liapunov) stable (with respect to the future) if, for any solution $t \to z(t)$ with $|z(0) - z_0(0)|$ sufficiently small, $|z(t) - z_0(t)|$ is arbitrarily small, uniformly for all instants $t \geq 0$. Stability with respect to the past is defined in an analogous fashion. (It is assumed that the solutions are defined for all $t \geq 0$ or $t \leq 0$, resp.)

12.2. Evidently the stability of an equilibrium solution is a special case of stability. But conversely, the general stability problem can be reduced to that of an equilibrium solution:

Using the above notation, let

$$y(t) = z(t) - z_0(t). \qquad \text{(Def.)}$$

Then the function y satisfies the differential equation

$$\frac{dy}{dt} = g(z_0(t) + y, t) - g(z_0(t), t),$$

which admits the equilibrium solution 0. The solution z_0 of the original differential equation is stable if, and only if, the equilibrium solution 0 of the new differential equation is stable.

Without loss of generality the discussion can therefore be restricted to the stability of an equilibrium solution 0.

12.3. Suppose that the autonomous Hamilton equation

$$\frac{dz}{dt} = JH_z(z)$$

has the equilibrium solution 0. Hence,

$$H_z(0) = 0$$

and, since without loss of generality it may be assumed that H(0) = 0,

$$H(z) = \tfrac{1}{2}\, z^* \cdot H_{zz^*}(0) \cdot z + 0(|z|^3) \qquad \text{as } |z| \to 0 .$$

Now assume that the Hamilton function H has a proper relative minimum H(0) = 0 at the point z = 0, and let ε be any positive real number sufficiently small. Then the continuous function H is positive at all point z with $0 < |z| < \varepsilon$, and it has a positive minimum, M (say), on the sphere of radius ε about the origin,

$$S_\varepsilon \underset{\text{Def.}}{=} \{z \,|\, |z| = \varepsilon\} .$$

There is a positive real δ such that

$$0 \leq H(z) < M \qquad \text{if } |z| < \delta .$$

By 10.1, H is a first integral of the Hamilton equation; that is,

$$H(z(t)) = H(z(0))$$

for any solution $t \to z(t)$ and any value of t. Therefore,

$$H(z(0)) < M$$

implies that

$$H(z(t)) < M \qquad \text{for all values of } t.$$

In particular, if $|z(0)| < \delta$, then

$$H(z(0)) < M,$$

and consequently

$$H(z(t)) < M.$$

Thus no point z(t) lies on the sphere S_ε, and since the starting point z(0) of the curve z(t) lies in the interior of S_ε, so does its whole range, the connected point set $\{z(t)|t \geq 0)$. We summarize:

If the Hamilton function H has a proper relative minimum at 0, in particular, if all the eigenvalues of the symmetric matrix $H_{zz*}(0)$ are positive, then the equilibrium solution 0 is Liapunov - stable with respect to both the past and the future.

12.4. In the preceding discussion it is irrelavant that the function the minimum of which has been used is just the Hamilton function of the system: relevant is only that the function is a first integral of the differential equation.

Further, that this differential equation is of the Hamiltonian type is also inessential: the same arguments apply to any differential equation

$$\frac{dz}{dt} = g(z).$$

Thus, by an argument analogous to that of §12.2, Dirichlet's criterion of stability is obtained:

Suppose that the (vector) differential equation

$$\frac{dz}{dt} = g(z)$$

has the equilibrium solution 0, and that there is a time-independent local first integral F(z) of the equation which has a proper relative minimum or maximum at z = 0. Then the equilibrium solution is Liapunov - stable.

The case of a maximum is reduced to that of a minimum by considering (-F) instead of F.

12.5. But how can we know whether or not a given function F is a first integral of the differential equation dz/dt = g(z)? If the definition of the notion of a first integral were to be applied, one would have explicitly to know the general solution of the differential equation; but then there would hardly be any need to look for first integrals!

Actually it is not necessary to know the general solution: by §10.8, a given non-constant function $z \to F(z)$ is a first integral of the differential equation

$$\frac{dz}{dt} = g(z),$$

if, and only if, it satisfies Liouville's equation

$$F_{z*}(z) \cdot g(z) = 0$$

for all points z in the domain considered.

12.6. The condition that F is an integral is unnecessarily restrictive. For any function $z \to F(z)$, whether or not it is a first integral, and any solution z(t) of the differential equation

$$\frac{dz}{dt} = g(z),$$

we have

$$\frac{d}{dt} F(z(t)) = F_{z*}(z(t)) \cdot g(z(t)).$$

In the argument of §§12.2-3 it was assumed that $\frac{d}{dt} F(z(t)) = 0$; but evidently it suffices there to suppose that

$$\frac{d}{dt} F(z(t)) \leq 0 \qquad (t \geq 0)$$

for all solution functions $t \to z(t)$ with starting points $z(0)$ sufficiently close to 0.

This condition may, in a straight-forward fashion, be further generalized to the case that the right-hand side of the differential equation, or the function F, or both, depend on the time variable t. Thus the following theorem, due to Liapunov, is obtained:

Let 0 be an equilibrium solution of the differential equation

$$\frac{dz}{dt} = g(z,t),$$

and $F(z,t)$ *be a real function of* z *and* t, *defined in some neighbourhood of the point* $z = 0$ *and for all* $t \geq 0$, *which satisfies the following conditions:*

(i) $F(0,t) = 0$ *for all* $t \geq 0$;

(ii) $\frac{\partial}{\partial t} F(z,t) + F_{z*}(z,t)g(z,t) \leq 0$

for all $t \geq 0$ *and all* z *in some open neighbourhood of* $z = 0$.

(iii) *there is a real function* $F_0(z)$, *which is defined in some open neighbourhood of* $z = 0$ *and has a proper relative minimum at* $z = 0$, *such that*

$$F(z,t) \geq F_0(z)$$

for all $t \geq 0$ *and all* z *in an open neighbourhood of* $z = 0$.

Then the equilibrium solution 0 is Liapunov-stable with respect to the future.

There is a similar proposition referring to the past.

12.7. Stability problems are often discussed by the method of small oscillations:

Let

$$\frac{dz}{dt} = g(z)$$

be an autonomous (vector) differential equation, with

$$g(0) = 0,$$

so that 0 is an equilibrium solution. Approximate, for sufficiently small $|z|$, $g(z)$ by $g_{z*}(0)z$, and the above differential equation by its linearization

$$\frac{dz}{dt} = Gz$$

where

$$G \underset{\text{Def.}}{=} g_{z*}(0).$$

(Jacobi's "equation of variation".) There are particular solutions of the form

$$z = ce^{\lambda t}$$

where c is some non-zero n-dimensional vector, and λ is a complex number. Such a function is a solution if, and only if,

$$(G - \lambda I_n)c = 0$$

and, consequently,

$$\det(G - \lambda I_n) = 0.$$

Suppose, for simplicity, that all the roots of this algebraic equation of the order n (the eigenvalues of the matrix G) are distinct. Then the n corresponding "eigenvectors" c are linearly independent, and the general solution of the linear differential equation is a linear combination of these n particular solutions with arbitrary coefficients. However, only such coefficients are physically meaningful which yield real-valued solutions. The following proposition is now obvious:

Suppose that the eigenvalues of the $(n\times n)$ matrix G are distinct. Then the equilibrium solution 0 of the linear differential equation

$$\frac{dz}{dt} = Gz$$

is Liapunov - stable with respect to the future if, and only if, all the real parts of the eigenvalues are non-positive.

Without a proof it may be noted that, if there are multiple eigenvalues, the necessary and sufficient condition for stability is that their real parts are negative, and those of the simple eigenvalues are non-positive.

12.8. For an application, suppose that for a given time-independent Hamilton function H with $H_z(0) = 0$, all the eigenvalues of the matrix

$$A \underset{\text{Def.}}{=} H_{zz*}(0)$$

are positive. Therefore, by §12.3, the equilibrium solution 0 both of the Hamilton equation

$$\frac{dz}{dt} = JH_z(z)$$

and its Jacobi equation

$$\frac{dz}{dt} = JAz$$

are Liapunov - stable. Hence, by §12.7, all the eigenvalues of the matrix JA have non-positive real parts. But the equilibrium solution 0 is stable also with respect to the past. It follows that those real parts must vanish:

Let A be any positive definite symmetric $(2m\times 2m)$ matrix. Then all the eigenvalues of the matrix JA are purely imaginary. With λ also $(-\lambda)$ is an eigenvalue of JA.

(For, if λ is an eigenvalue, so is its complex conjugate $-\lambda$.)

12.9. The method of 12.7 is restricted to differential equations

$$\frac{dz}{dt} = g(z,t)$$

whose equations of variation

$$\frac{dz}{dt} = g_{z*}(0,t)\ z$$

are autonomous, i.e. for which the matrices $g_{z*}(0,t)$ are time-independent. Liapunov's stability theorem, 12.6, however, can also be applied to non-autonomous linear equations

$$\frac{dz}{dt} = G(t)z.$$

Let

$$F(z) = \tfrac{1}{2}\ z^*Cz$$

where C denotes a positive-definite symmetric matrix. Then the condition of Proposition 12.6 is satisfied if the quadratic form $\frac{1}{2}\ x^*B(t)x$, where

$$B(t) \underset{\text{Def.}}{=} CG(t) + G^*(t)C\ ,$$

is non-positive for all t sufficiently large, i.e. if all the eigenvalues of B(t) (which are real) are non-positive for all large t. This holds, in particular, when C is the (n×n) unit matrix, I_n.

12.10. This last result is also obtained by a slightly different method: Let, for $z \neq 0$,

$$\rho = |z|$$

and

$$u = z/|z|,$$

so that the differential equation

$$\frac{dz}{dt} = G(t)z$$

is transformed into the equivalent equation

$$\frac{d\rho}{dt}\ u + \rho\ \frac{du}{dt} = \rho G(t)u.$$

Hence, by a scalar multiplication with the unit vector u,

$$\frac{d\rho}{dt} = \rho \cdot u^*G(t)u,$$

and

$$\rho = \rho_0 \cdot \exp\left\{ \int_0^t u^*G(\tau)ud\tau \right\}$$

with $\rho = \rho_0$ for $t = 0$. Hence the proposition:

Let $\lambda'(t)$ and $\lambda''(t)$ denote the least and the largest eigenvalue of the quadratic form $u^* \cdot \frac{1}{2}(G(t) + G(t))u$ at the time t, and z(t) be any solution of the linear differential equation

$$\frac{dz}{dt} = G(t)z.$$

Then, for any $t \geq 0$,

$$|z(0)|\exp\left\{ \int_0^t \lambda'(\tau)d\tau \right\} \leq |z(t)| \leq |z(0)|\exp\left\{ \int_0^t \lambda''(\tau)d\tau \right\} .$$

In particular, if $\int_0^t \lambda''(\tau)d\tau$ is bounded, then the equilibrium solution 0 is stable, and if this integral tends to $-\infty$ as $t \to \infty$, then $z(t) \to 0$ as $t \to \infty$.

12.11. For an application, suppose that the matrix G is independent of the time-variable t; let α' and α'' be the minimum and the maximum of the real parts of the eigenvalues of G, and λ' and λ'' be the least and the greatest eigenvalue of the symmetric matrix $\frac{1}{2}(G + G^*)$. Then,

$$\lambda' \leq \alpha' \leq \alpha'' \leq \lambda''.$$

Proof. A straight-forward computation shows that for any eigensolution z(t) of the equation $dz/dt = G\,z$ (i.e. for any solution of the form $z(0) \cdot e^{\gamma t}$ where γ is an eigenvalue of the matrix G)

$$|z(t)| = |z(0)| \cdot e^{\alpha t} \cdot \xi(t);$$

here α denotes the real part of the complex number γ, and ξ is a bounded function, depending on γ. An application of Proposition 12.10 now yields the assertion.

12.12. The stability of the solution 0 of the equation of variation,

$$\frac{dz}{dt} = g_{z*}(0,t)z,$$

does not necessarily imply the stability of the solution 0 of the non-linear equation

$$\frac{dz}{dt} = g(z,t).$$

For a counter-example see the literature. A sufficient condition is given by the following proposition:

Suppose that

$$g(z,t) = G(t)z + q(z,t)$$

for all $t \geq 0$, and all z sufficiently close to 0; that the eigenvalues of all the matrices $\frac{1}{2}(G(t) + G^*(t))$, $t \geq 0$, are bounded above by the negative real number $(-\Lambda)$; and that, for any positive real $\varepsilon < \Lambda$, there is a positive real δ such that

$$|q(z,t)| < \varepsilon \cdot |z| \quad \text{for any} \quad t \geq 0 \quad \text{and} \quad |z| < \delta .$$

Let $z(t)$ be a solution of the equation

$$\frac{dz}{dt} = g(z,t)$$

such that $|z(0)| < \delta$. Then, for any $t \geq 0$, $|z(t)| \leq |z(0)|\, e^{-(\Lambda - \varepsilon)t}$.

Proof. Using polar coordinates (cf. §12.10) yields, for $\rho = |z|$,

$$\frac{d \log \rho}{dt} \leq -\Lambda + \varepsilon ,$$

provided that $|\rho(t)| < \delta$. This inequality holds for all $t \geq 0$ if it holds for $t = 0$. Hence the assertion by "continuous induction".

12.13. The method of polar coordinates of 12.10 can be generalized, in a straight-forward manner, to differential equations

$$\frac{dz}{dt} = G(z,t)$$

whose right-hand sides are homogeneous in z, and there is a corresponding stability theorem which is analogous to Proposition 12.12.

Suppose that G is homogeneous in z of the order $\alpha > 1$, and the function $u*G(u,t)$ is on the unit sphere $|u| = 1$ bounded above by the non-positive number $-\Gamma$, uniformly for all $t \geq 0$. Then, the equilibrium solution 0 of the equation

$$\frac{dz}{dt} = G(z,t)$$

is stable. For any solution z(t) of this equation,

$$|z(t)|^{\alpha-1} \leq \frac{1}{|z(0)|^{1-\alpha}+(\alpha-1)\Gamma t}, \quad t \geq 0 .$$

12.14. The generalization of Proposition 12.12 now reads:

In addition to the assumption of Proposition 12.13, suppose that $-\Gamma$ is negative, that

$$g(z,t) = G(z,t) + q(z,t), \quad t \geq 0,$$

and that, for any positive real $\varepsilon < \Gamma$, there is a positive real δ such that

$$|q(z,t)| < \varepsilon \cdot |z|^{\alpha} \quad \text{for all } t \geq 0 \text{ and } |z| < \delta$$

Then, for any solution z(t) of the equation

$$\frac{dz}{dt} = g(z,t)$$

with $|z(0)| < \delta$,

$$|z(t)|^{\alpha-1} \leq \frac{1}{|z(0)|^{1-\alpha}+(\alpha-1)(\Gamma-\varepsilon)t} \quad \text{for} \quad t \geq 0.$$

References and Notes

§12.1. The term "stability" has several meanings: see, for example, Lefschetz (1957) or Nemytskii - Stepanov (1960). In the text only one of them is discussed.

§§12.3-6. A few historical remarks concerning the development of the stability criterion from Lagrange via Dirichlet to Liapunov are found in Lefschetz (1957). Liapunov's criterion can be further extended even to non-differentiable functions F. More details are found e.g. in Birkhoff (1927), Bellman (1953), Lefschetz (1957), Nemytskii-Stepanov (1960), Hahn (1963), (1967), Wilson (1971).

§12.12. For a counter-example see, for example, Wintner (1952), §136.

CHAPTER IV

JACOBI'S GEOMETRIC INTERPRETATION OF DYNAMICS

§13. Maupertuis' Principle

13.1. In this chapter only conservative systems, i.e. systems with time-independent Lagrange functions and, therefore, time-independent Hamilton functions will be considered. The Hamilton function H is, by §10.1, a first integral of the Hamilton equations. There is a corresponding first integral of the Lagrange equations with Lagrange function $L(x,\dot{x})$, which we denote by $Q(x,\dot{x})$:

The function

$$Q(x,\dot{x}) \underset{\text{Def.}}{=} \dot{x}^* \cdot L_{\dot{x}}(x,\dot{x}) - L(x,\dot{x})$$

is a first integral of the Lagrange equations with time-independent Lagrange function $L(x,\dot{x})$.

For, by definition of Q,

$$Q(x,\dot{x}) = H(x,y),$$

provided that the canonical coordinates (x,y) and the Lagrangian coordinates correspond to each other by the Legendre transformation of §4.2, implicitly defined by means of the equation

$$H(x,y) + L(x,\dot{x}) = \dot{x}^* \cdot y.$$

Hence the assertion.

Alternatively the lemma can be proven by a straight-forward verification that the function Q satisfies the Liouville equation corresponding to the Lagrange equations considered.

13.2. By Hamilton's Principle 9.2,

$$\left[\frac{\partial}{\partial\varepsilon}\int_{t^0}^{t^1} L(x + \varepsilon\xi, \frac{dx}{dt} + \varepsilon\frac{d\xi}{dt})dt\right]_{\varepsilon=0} = 0$$

for any Lagrange curve x(t) (any solution x(t) of Lagrange's equations) and an arbitrary admissible function $\xi(t)$. As to admissibility see §9.1. Now express L in terms of $L_{\dot{x}}$ and Q, using Lemma 13.1, and subject the functions ξ to the restriction that

$$Q(x+\varepsilon\xi, \frac{dx}{dt} + \varepsilon\frac{d\xi}{dt}) - Q(x, \frac{dx}{dt}) = O(\varepsilon^2) \text{ as } \varepsilon \to 0.$$

That is, now only such curves $y(t) = x(t) + \varepsilon\cdot\xi(t)$ are admitted which, in a sense, are close to the "energy surface" in the $(x,\dot{x})$-space with the equation

$$Q(x,\dot{x}) = h,$$

where h denotes the constant value of the first integral Q for the Lagrange curve x(t). This condition is satisfied if, and only if,

$$Q^*_x(x, \frac{dx}{dt}) \cdot \xi + Q^*_{\dot{x}}(x, \frac{dx}{dt}) \cdot \frac{d\xi}{dt} = 0$$

for all admitted values of t.

Carrying out the operations mentioned yields the following proposition:

<u>Let</u> x(t) <u>be any solution of the Lagrange equations with the time-independent Lagrange function</u> $L(x,\dot{x})$, h <u>be the corresponding value of the first integral</u> Q <u>introduced in §13.1</u>:

$$h = Q(x, \frac{dx}{dt}),$$

<u>and</u> $\xi(t)$ <u>be any admissible vector function such that for all sufficiently small</u> ε <u>the curve</u> $y(t) = x(t) + \varepsilon\cdot\xi(t)$ <u>is close to the energy surface</u> $\{(x,\dot{x})\,|\,Q(x,\dot{x}) = h\}$ <u>in the sense explained</u>. <u>Then, for any two reals</u>

t^0 *and* t^1

$$\left[\frac{\partial}{\partial\varepsilon}\int_{t^0}^{t^1}\frac{dy^*}{dt}L_{\dot{x}}\left(y,\frac{dy}{dt}\right)dt\right]_{\varepsilon=0} = 0.$$

That is, under the restrictions made concerning the functions ξ , the first variation of the functional

$$\int_{t^0}^{t^1}\frac{dy^*}{dt}\cdot L_{\dot{y}^*}\left(y,\frac{dy}{dt}\right)dt$$

vanishes for $y(t) = x(t)$).

13.3. *Corollary* (Maupertuis' Principle). *Suppose that*

$$L(x,\dot{x}) = G(x,\dot{x}) - V(x),$$

where G *is a homogeneous function of* $\dot{x}$ *of the positive order* α; *that is,*

$$G(x,\lambda\dot{x}) = \lambda^{\alpha}\cdot G(x,\dot{x})$$

for any positive real λ. *Then,*

$$\left[\frac{\partial}{\partial\varepsilon}\int_{t^0}^{t^1}G\left(y,\frac{dy}{dt}\right)dt\right]_{\varepsilon=0} = 0.$$

Proof.

$$\dot{x}^*L_{\dot{x}} = \dot{x}^*G_{\dot{x}}$$

$$= \alpha G, \quad \text{by Euler's identity.}$$

Hence, by Proposition 13.2, the assertion.

13.4. In the following, curves y(t) will be considered which are not only close to the energy surface $\{(x,\dot{x}) | Q(x,\dot{x}) = h\}$ (in the sense of §13.2), but actually lie in it. The curves $x(t) + \varepsilon\xi(t)$, with (sufficiently small) arbitrary ε, will in general not lie in that surface. However, functions $\xi(t,\varepsilon)$, with $\xi(t,0) \neq 0$, can be determined in such a way that, for all ε sufficiently close to 0,

$$Q(x + \varepsilon\xi, \frac{dx}{dt} + \varepsilon \frac{d\xi}{dt}) - Q(x, \frac{dx}{dt}) = 0.$$

In a straight-forward fashion it is verified that the conclusions of Proposition 13.2. and of Maupertuis' Principle 13.3 hold also in this case.

Note

§13.3. In the literature it is often assumed that $G(x,\dot{x})$ is a positive definite quadratic form in $\dot{x}$. As the text shows, this restriction is not necessary.

§14. Riemannian Geometry

14.1. Maupertuis' Principle (§13.3) applies, in particular, to dynamical systems with Lagrange functions of the form

$$L(x,\dot{x}) = \tfrac{1}{2}\, \dot{x}^* G(x)\dot{x} - V(x)$$

where G denotes an $(m\times m)$-matrix function of the position vector x. The corresponding "Lagrange motions", i.e. the solutions $t \to x(t)$ of the corresponding Lagrange equations can be interpreted in geometric terms, as Jacobi has shown. The present section is to explain those concepts and results of differential geometry which are needed for the formulation and derivation of Jacobi's interpretation.

14.2. A subset of the 3-dimensional Euclidean space E^3 which is a 2-dimensional analytic manifold will be called a surface in E^3. Let x be a point of a given surface, having the Cartesian coordinates x^1, x^2, x^3, and $u \to x$, with $u = (u^1, u^2) \in E^2$, be a local parametrization of the surface. Then the rank of the matrix $\{\partial x^\alpha / \partial u^i\}$ is 2 ; hence, the vectors x_1 and x_2, defined by

$$x_i = \partial x / \partial u^i, \; i = 1,2,$$

are not collinear, the vector product $x_1 \times x_2$ does not vanish. (We have here identified the point x with its position vector (x^1, x^2, x^3). A surface which is covered by a single system of local coordinates will be called a local surface. Thus a surface is the union of at most countably many local surfaces.

14.3. Let $t \to f(t)$ be a curve on the surface, and α and $\beta > \alpha$ be any two admitted values of the parameter t. Then the (Euclidean) length of the arc from $f(\alpha)$ to $f(\beta)$ is

$$\int_\alpha^\beta \sqrt{\frac{df^*}{dt} \frac{df}{dt}} \, dt$$

Now suppose that the arc belongs to one coordinate neighbourhood on the surface. Then to the arc $t \to f(t)$ on the surface there corresponds an arc $t \to u(t)$, $\alpha \le t \le \beta$, in the parameter plane E^2 such that

$$x(u(t)) = f(t), \; \alpha \le t \le \beta.$$

Hence the length of the arc f on the surface is also given by

$$\int_\alpha^\beta \sqrt{g_{ij}(u) \frac{du^i}{dt} \frac{du^j}{dt}} \, dt$$

where

$$g_{ij}(u) \underset{\text{Def.}}{=} x_i^*(u)x_j(u) \ , \ i = 1,2,$$

and the summation convention has been applied. The matrix

$$g(u) \underset{\text{Def.}}{=} \{g_{ij}(u)\}_{i,j=1}^{2}$$

is, by its definition, positive definite.

<u>14.4</u>. The above integral yielding the length of arc is invariant under one-to-one transformations of the parameter t. Therefore it may be written as

$$\int_{\alpha}^{\beta} \sqrt{g_{ij}(u)du^i \ du^j}$$

where a and b denote the starting and end points in the u-plane of the arc considered. Thus, although the point-set curve connecting a and b is fixed, its parametrization is left open. Often the length of arc, s, from some fixed point on the curve to the variable point, defined by

$$s = \int_{t^0}^{t} \sqrt{g_{ij}(u) \frac{du^i}{dt} \frac{du^j}{dt}} \ dt,$$

is used as the parameter of the curve.

<u>14.5</u>. When the "Gaussian parameters" u^1, u^2 are subject to a transformation $u \to \bar{u}$ of local coordinates, the "<u>first fundamental form</u>" of the surface, $g_{ij}(u)du^i \ du^j$, is transformed to

$$\bar{g}_{k\ell}(\bar{u})d\bar{u}^k \ d\bar{u}^\ell = g_{ij}(u) \frac{\partial u^i}{\partial \bar{u}^k} \frac{\partial u^j}{\partial \bar{u}^1} \ du^i \ du^j \ .$$

Hence, <u>both the matrix functions</u>

$$g(u) = \{g_{ij}(u)\}_{i,j=1}^{2}$$

and

$$\bar{g}(\bar{u}) = \{\bar{g}_{ij}(\bar{u})\}_{i,j=1}^{2}$$

are representatives of a covariant tensor field defined on the surface. (See §3.20.) It is called the field of the metric or first fundamental tensor of the surface. Any matrix responding a tensor of this field is positive definite: thus we shall say the field of the metric tensor is positive definite. If confusion appears unlikely, the distinction between a tensor field and its representative (with respect to a given system of coordinates) will occasionally be ignored in our notation and terminology.

14.6. Although a surface is not uniquely determined by its metric tensor field, some of its geometric properties are so, e.g. its "Gaussian curvature". Thus it is reasonable to investigate those properties of a surface which depend only on its metric tensor field. The theory thus obtained is the "intrinsic geometry" of the surface. The intrinsic geometry of a single coordinate neighbourhood of the surface is called its local intrinsic geometry. The local instrinsic geometries of a plane, a cylinder, and a cone are identical although these surfaces, when considered as embedded in E^3, are essentially different.

Observations like this one suggest us to introduce the notion of an abstract two-dimensional Riemann manifold: it is defined as a connected two-dimensional analytic manifold on which a positive definite covariant tensor field g of the second order is defined, which takes the role of a metric tensor field. That is, if u^1, u^2 are local coordinates in a neighbourhood of a point of the manifold, and $t \to (u^1,u^2)$ is a differentiable curve in an appropriate neighbourhood in the u-plane, then the number

$$\int_{\alpha}^{\beta} \sqrt{g_{ij}(u) \frac{du^i}{dt} \frac{du^j}{dt}}\, dt$$

is assigned to the corresponding arc $t \to x(u(t))$ on the manifold as its length.

Note that here nothing has been said about an embedding of the manifold in E^3 or any other Euclidean space E^k. In particular, it has not been supposed that

$$g_{ij}(u) = \left(\frac{\partial x}{\partial u^i}\right)^* \frac{\partial x}{\partial u^j}$$

for some mapping $u \to x \in E^k$.

14.7. The generalization to m dimensions is straight-forward: an m-dimensional Riemann manifold is defined as an m-dimensional connected analytic manifold on which a metric tensor field $g = \{g_{ij}\}$ is given, i.e. a positive definite covariant tensor field of the second order, which yields the length of any differentiable arc by the expression introduced in §14.6.

14.8. Let a and b two points of a Riemann manifold which lie in one coordinate neighbourhood, and $s \to u(s)$ be an arc which lies entirely in this neighbourhood, connects a and b, and is of minimal length; that is, the length of any other differentiable arc connecting a and b and situated in the neighbourhood is not less than the length of u(s). Then the function u(s) satisfies the differential equation

$$\frac{d^2u^k}{ds^2} + \Gamma^k_{ij} \frac{du^i}{ds} \frac{du^j}{ds} = 0$$

<u>where</u>

$$\left.\begin{aligned} \Gamma^k_{ij}(u) &= g^{k\ell}(u)\Gamma_{ij,\ell}(u) , \\ \{g^{k\ell}(u)\}^m_{k,\ell=1} &= (g(u))^{-1} , \\ \Gamma_{ij,\ell} &= \tfrac{1}{2}(-\frac{\partial g_{ij}}{\partial u^\ell} + \frac{\partial g_{j\ell}}{\partial u^i} + \frac{\partial g_{\ell i}}{\partial u^j}). \end{aligned}\right\} \text{(Def.)}$$

<u>The parameter</u> s <u>is the length of arc of the curve</u> u.

The symbols "$\Gamma_{ij,\ell}$" and "Γ^k_{ij}" are called the <u>Christoffel symbols</u> of the first and second kind, resp.; in general, they do <u>not</u> denote representatives of tensor fields.

<u>Proof</u>. Suppose that first the minimizing arc is given as a function $t \to u(t)$ of a parameter t which need not be the length of arc, and that

$$u(\alpha) = a, \ \mu(\beta) = b.$$

Determine, by §9.1, the first variation of the functional

$$\int_\alpha^\beta \sqrt{g_{ij}(u + \varepsilon\xi) \frac{d}{dt} (u^i + \varepsilon\xi^i) \cdot \frac{d}{dt} (u^j + \varepsilon u^j)} \ dt,$$

and then choose $t = s$, so that

$$g_{ij}(u) \frac{du^i}{ds} \frac{du^j}{ds} = 1.$$

The corresponding Euler equations read

$$\tfrac{1}{2} \frac{\partial g_{ij}}{\partial u^k} \frac{du^i}{ds} \frac{du^j}{ds} - \frac{d}{ds} (g_{ik} \frac{du^i}{ds}) = 0$$

or equivalently, by using the Christoffel symbols of the first kind,

$$g_{i\ell} \frac{d^2u^i}{ds^2} + \Gamma_{ij,\ell} \frac{du^i}{ds} \frac{du^j}{ds} = 0,$$

and finally,

$$\frac{d^2u^k}{ds^2} + \Gamma^k_{ij} \frac{du^i}{ds} \frac{du^j}{ds} = 0,$$

as has been asserted.

Any curve $s \to u(s)$ which satisfies this differential equation is called a <u>geodesic</u> on the Riemann manifold. Thus any arc of minimal length is a geodesic. The converse is generally true only if the end-points of the arc are sufficiently close to each other. (Cf. §14.13.)

<u>14.9</u>. Euclidean geometry is a Riemann geometry the metric tensors of which, in suitable systems of local coordinates, are represented by the $(m \times m)$ unit matrix I_m. If $g(u) = I_m$, all the Christoffel symbols vanish identically:

$$\Gamma_{ij,k}(u) = 0, \ \Gamma^k_{ij}(u) = 0,$$

and the equation of the geodesic reads:

$$\frac{d^2u^k}{ds^2} = 0;$$

that is, the geodesics are the Euclidean straight lines.

<u>14.10</u>. A Euclidean straight line is determined by any of its points and by its direction, described by a unit vector. The same holds for geodesics in general: by the existence and uniqueness theorems concerning ordinary differential equations (§2.9), any solution ϕ of the second-order differential equation for geodesics is uniquely determined by any of its points, $\overset{o}{u}$ (say) and the direction vector (i.e. tangent vector) a at this point.

<u>14.11</u>. A Euclidean straight line is also uniquely determined by any two of its points. There is the following partial generalization of this theorem:

<u>For any point</u> $\overset{o}{u}$ <u>of a Riemann manifold there is a neighbourhood such that any two of its points can be connected by exactly one geodesic arc</u>.

This can be proven by means of the following lemmas:

14.12. Lemma. Let $\overset{o}{u}$ be a point of the Riemann manifold, and a be a vector such that

$$||a|| \underset{\text{Def.}}{=} +\sqrt{a^* g(\overset{o}{u}) a} = 1.$$

Then, for some positive real number ε depending only on $\overset{o}{u}$, there is exactly one solution

$$u = \phi(\overset{o}{u}, a, s)$$

of the differential equation of §14.8 for geodesics which start at the point $\overset{o}{u}$ with the initial tangent vector a and is defined for all $s \in [0,\varepsilon]$.

The lemma can be proven indirectly, by using the fact that the "unit sphere" $\{a \in E^m | \quad ||a|| = 1\}$ is a bounded and closed subset of E^m.

14.13. Lemma. Assume the hypotheses and notation of the preceding lemma, and let λ be any positive real ≤ 1. Then

$$\phi(\overset{o}{u}, \lambda a, s) = \phi(\overset{o}{u}, a, \lambda s) \text{ for } 0 \leq s \leq \varepsilon.$$

Proof. Both sides of the asserted equation satisfy the differential equation of geodesics (§14.8), and both take, at $s = 0$, the initial value λa. Hence, by the existence and uniqueness theorem of §2.9, they are equal.

14.14. Lemma. With the same hypotheses and notation,

$$\phi(\overset{o}{u}, \frac{s}{\varepsilon} a, \varepsilon) = \phi(\overset{o}{u}, a, s) \qquad \text{for } 0 \leq s \leq \varepsilon \ .$$

Proof. Substitute, in Lemma 14.13, $\frac{s}{\varepsilon}$ for λ, and ε for s.

14.15. Lemma. With the same hypotheses and notation, the Jacobian determinant of the mapping $c \to u$, defined by

$$u = \phi(\overset{\circ}{u}, c, \varepsilon),$$

is, in some neighbourhood of the point $c = 0$, positive.

Proof. Differentiate the equality

$$\phi(\overset{\circ}{u}, c\,\frac{s}{\varepsilon}, \varepsilon) = \phi(\overset{\circ}{u}, c \not\equiv, s)$$

with respect to s at $s = 0$; hence,

$$\phi_{c*}(\overset{\circ}{u}, 0, \varepsilon) = \varepsilon\, I_m,$$

which implies the assertion.

14.16. Lemma. There is an open neighbourhood C of $c = 0$ and an open neighbourhood $\overset{\circ}{U}$ of $\overset{\circ}{u}$ such that the mapping $\phi : C \to \overset{\circ}{U}$ is one-to-one.

The assertion is an immediate consequence of the previous lemma and the inverse-function theorem.

The components of the vector $c \in C$ are said to be the Riemann normal coordinates of the point $u = \phi(\overset{\circ}{u}, c, \varepsilon) \in \overset{\circ}{U}$.

14.17. Now a special case of Proposition 14.11 can be proven:
For any point $\overset{\circ}{u}$ of a Riemann manifold there is an open neighbourhood such that each of its points is connected with $\overset{\circ}{u}$ by exactly one geodesic arc.

Proof. Let u be an arbitrary point sufficiently close to $\overset{\circ}{u}$ so that, for some vector c,

$$\phi(\overset{\circ}{u}, c, \varepsilon) = u. \qquad \text{(Cf. §14.16.)}$$

Then the geodesic arc

$$\phi(u, \frac{c}{\|c\|}, s) \qquad 0 \le s \le \varepsilon\, \|c\|,$$

connects the points $\overset{\circ}{u}$ and u. Its uniqueness is shown by a straight-forward indirect argument.

14.18. Proof of Proposition 14.11: The mapping

$$\begin{cases} \bar{u} = u, \\ \bar{v} = \phi(u,v,\varepsilon), \end{cases}$$

defined in some neighbourhood of the point $\binom{\mathring{u}}{0} \in E^m \times E^m$, has the functional matrix

$$\begin{pmatrix} I_m & 0 \\ \phi_{u*} & \phi_{v*} \end{pmatrix}$$

which, by §14.15, has a positive determinant at $\binom{\mathring{u}}{0}$. Hence, by the inverse function theorem, there is a neighbourhood of $\binom{\mathring{u}}{0}$ in which the mapping is one-to-one. Therefore in certain open neighbourhoods P and Q of $\mathring{u}$ and 0, there are, for each $\bar{u} \in P$, $\bar{v} \in Q$, exactly one u and one v such that the above transformation equations hold. Thus for each point u in P, and each point whose vector $\bar{v}$ of Riemann normal coordinates lies in Q, there is exactly one vector v such that

$$\bar{v} = \phi(u,v,\varepsilon).$$

Thus both those points are connected by the geodesic arc

$$t \to \phi(u,v,\varepsilon t), \quad 0 \leq t \leq 1 \ .$$

14.19. Geodesic arcs have the following minimal property:

Let $\mathring{u}$ be a point of a Riemann manifold M, and u be a point of M sufficiently close to $\mathring{u}$. Then the length of the geodesic arc connecting $\mathring{u}$ and u (cf. §14.12) is less than the length of any other arc connecting $\mathring{u}$ and u.

For the proof, let $r \to \phi(\mathring{u},a,r)$, with $\|a\| = 1$ and $r \geq 0$, be any geodesic arc starting at $\mathring{u}$. Introduce r and a as polar coordinates. With respect to these coordinates,

$$ds^2 = dr^2 + \sum_{i,j=1}^{m-1} g_{ij}(u)du^i du^j,$$

which implies the assertion. For the details see the literature.

References

Laugwitz (1965), Stoker (1969), Milnor (1969). In particular, Milnor gives a complete proof of Proposition 14.19.

§15. Jacobi's Geometric Interpretation of Conservative Lagrangian Systems

15.1. In this section, as in §14.1, a Lagrangian system will be considered whose local Lagrange function L is given by

$$L(x,\dot{x}) = \tfrac{1}{2}\, \dot{x}^* g(x)\dot{x} - V(x);$$

where g(x) denotes a positive definite (m×m) matrix. By §§13.3 and 13.4

$$\left[\frac{\partial}{\partial\varepsilon} \int_{t^0}^{t^1} \tfrac{1}{2} \left(\frac{dx}{dt} + \varepsilon \frac{d\xi}{dt}\right)^* \cdot g(x + \varepsilon\xi) \cdot \left(\frac{dx}{dt} + \varepsilon \frac{d\xi}{dt}\right)dt\right]_{\varepsilon=0} = 0$$

where $t \to x(t)$ denotes any given Lagrange curve with $dx/dt \neq 0$ for all t considered, and $t \to x(t) + \varepsilon\cdot\xi(t,\varepsilon)$ is a curve on or close to the energy surface

$$E = \{(x,\dot{x}) | Q(x,\dot{x}) = h\}$$

("close to" in the sense of §13.2).

Using the above notation, let $\tilde{M}$ *be the set of all those points* x *considered for which*

$$V(x) < h \quad ,$$

and define in $\tilde{M}$ *a Riemann metric* $\tilde{g}$ *by*

$$\tilde{g}(x) = 2(h - V(x)) \cdot g(x)$$

(called a "Jacobi metric".) Then,

$$\left[\frac{\partial}{\partial \varepsilon} \int_{t^0}^{t^1} \sqrt{\left(\frac{dx}{dt} + \varepsilon \frac{d\xi}{dt}\right)^* \tilde{g}(x + \varepsilon\xi)\left(\frac{dx}{dt} + \varepsilon \frac{d\xi}{dt}\right)}\, dt\right]_{\varepsilon=0} = 0.$$

Proof.

(1) Let $y(t) = x(t) + \varepsilon \cdot \xi(t,\varepsilon)$.

(2) $\frac{1}{2} \frac{dy^*}{dt} g(y) \frac{dy}{dt} = Q(y, \frac{dy}{dt}) - V(y)$ (cf. §13.1)

$= Q(x, \frac{dx}{dt}) + o(\varepsilon^2) - V(y)$

$= h - V(y) + o(\varepsilon^2)$ for $\varepsilon \to 0$. (Hyp.)

(3) $h - V(y) > 0$ for $|\varepsilon|$ sufficiently small.

(4) $\frac{1}{2} \frac{dy^*}{dt} g(y) \frac{dy}{dt} = \sqrt{h - V(y) + o(\varepsilon^2)}$

$= \sqrt{h - V(y)} + o(\varepsilon^2)$, $\varepsilon \to 0$.

(5) $\frac{1}{2} \frac{dy^*}{dt} g(y) \frac{dy}{dt} = \sqrt{\frac{1}{2} \frac{dy^*}{dt} g(y) \frac{dy}{dt}} \sqrt{\frac{1}{2} \frac{dy^*}{dt} g(y) \frac{dy}{dt}}$

$= \sqrt{h - V(y)} \sqrt{\frac{1}{2} \frac{dy^*}{dt} g(y) \frac{dy}{dt}} + o(\varepsilon^2)$.

Hence the assertion.

15.2. The condition just established for the Lagrange curve $t \to x(t)$ is formally identical with that for a geodesic as given in §14.8. But there is a difference: the functions ξ are, in the present case, subject to the conditions of §13.2 that

$$Q_{y^*} \xi + Q_{\dot{y}^*} \frac{d\xi}{dt} = 0,$$

whereas in §14.8 there has been no such restriction. Nevertheless, the equation of a geodesic holds also for the Lagrange curve, as will be shown in the following.

Lemma. The Lagrange equations corresponding to the Lagrange function

$$L(x,\dot{x}) = \tfrac{1}{2}\,\dot{x}^{*}g(x)\dot{x} - V(x)$$

are equivalent to the equations

$$\frac{d^2x^k}{dt^2} + \Gamma^k_{ij}\frac{dx^i}{dt}\frac{dx^j}{dt} + g^{k\ell}\frac{\partial V}{\partial x^\ell} = 0,\ k=1,2,\cdots,n,$$

where the Γ's are the Christoffel symbols of the second kind formed by means of the metric g(x).

(Again, two systems of differential equations are said to be equivalent if each solution of either system is also a solution of the other system.)

Proof. The Lagrange equations can be given the form

$$\left(g_{kj}\frac{d^2x^j}{dt^2} + \frac{\partial g_{ij}}{\partial x^k}\frac{dx^i}{dt}\frac{dx^j}{dt}\right) - \left(\tfrac{1}{2}\frac{\partial g_{ij}}{\partial x^k}\frac{dx^i}{dt}\frac{dx^j}{dt} - \frac{\partial V}{\partial x^k}\right) = 0.$$

Hence, by the definition of the Γ's,

$$g_{kj}\frac{d^2x^j}{dt^2} + \Gamma_{ij,k}\frac{dx^i}{dt}\frac{dx^j}{dt} + \frac{\partial V}{\partial x^k} = 0.$$

This implies the equation asserted in the lemma.

Conversely, from these equations the Lagrange equations are obtained by reversing the argument.

15.3. Let $t \to x(t)$ be a solution of the Lagrange equations for the local Lagrange function

$$\tfrac{1}{2}\,\dot{x}^{*}g(x)\dot{x} - V(x) \qquad \text{(cf. §15.1)},$$

$$h \underset{\text{Def.}}{=} Q\left(x, \frac{dx}{dt}\right) \quad \text{be its "energy"},$$

$$\tilde{g}(x) \underset{\text{Def.}}{=} 2(h - V(x))g(x)$$

be the corresponding Jacobi metric, the parameter $\tilde{s}$ be the length of arc with respect to the Jacobi metric, and

$$\tilde{x}(\tilde{s}) = x(t). \qquad \text{(Def.)}$$

Then,

$$\frac{d^2\tilde{x}^k}{d\tilde{s}^2} + \tilde{\Gamma}^k_{ij} \frac{d\tilde{x}^i}{d\tilde{s}} \frac{d\tilde{x}^j}{d\tilde{s}} = 0.$$

(The $\tilde{\Gamma}$'s are the Christoffel symbols of the second kind corresponding to the Jacobi metric $\tilde{g}$.) That is, any Lagrange curve is transformed, by a suitable parameter transformation, into a geodesic with respect to the appropriate Jacobi metric.

Proof.

(1) Make, in the equation

$$g_{kj} \frac{d^2x^j}{dt^2} + \Gamma_{ij,k} \frac{dx^i}{dt} \frac{dx^j}{dt} + \frac{\partial V}{\partial x^k} = 0$$

of Lemma 15.2, the substitutions

$$g_{ij} = \tilde{g}_{ij}/2(h-V) ,$$

$$\frac{dx^j}{dt} = \frac{d\tilde{s}}{dt} \frac{d\tilde{x}^j}{ds} ,$$

$$\frac{d^2x^j}{dt^2} = \frac{d^2\tilde{s}}{dt^2} \frac{d\tilde{x}^j}{d\tilde{s}} + \left(\frac{d\tilde{s}}{dt}\right)^2 \frac{d^2\tilde{x}^j}{d\tilde{s}^2} .$$

(2) Thus,

$$\tfrac{1}{2} \frac{\tilde{g}_{kj}}{h-V} \frac{d^2\tilde{s}}{dt^2} \frac{d\tilde{x}^j}{d\tilde{s}} + \tfrac{1}{2} \frac{\tilde{g}_{kj}}{h-V} \left(\frac{d\tilde{s}}{dt}\right)^2 \frac{d^2\tilde{x}^j}{d\tilde{s}^2}$$

$$+ \tfrac{1}{2} \frac{\tilde{\Gamma}_{ij,k}}{h-V} \left(\frac{d\tilde{s}}{dt}\right)^2 \frac{d\tilde{x}^i}{d\tilde{s}} \frac{d\tilde{x}^j}{d\tilde{s}}$$

$$+ \tfrac{1}{4}(h-V)^{-2}\left(-g_{ij}\frac{\partial V}{\partial x^k} + g_{jk}\frac{\partial V}{\partial x^i} + g_{ki}\frac{\partial V}{\partial x^j}\right)\left(\frac{d\tilde{s}}{dt}\right)^2 \frac{d\tilde{x}^i}{d\tilde{s}}\frac{dx^j}{d\tilde{s}} + \frac{\partial V}{\partial x^k} = 0.$$

(3) $$\frac{d\tilde{s}}{dt} = \sqrt{\tilde{g}_{ij}\frac{dx^i}{dt}\frac{dx^j}{dt}}$$

$$= \sqrt{2(h-V)g_{ij}\frac{dx^i}{dt}\frac{dx^j}{dt}}$$

$$= \sqrt{2(h-V)\cdot 2(h-V)}$$

$$= 2[h-V(x(t))]\ .$$

(4) $$\frac{d^2\tilde{s}}{dt^2} = \frac{d}{dt}\frac{d\tilde{s}}{dt}$$

$$= 2\frac{d}{dt}(h-V) \qquad (3)$$

$$= -2\frac{\partial V}{\partial\tilde{x}^\ell}\frac{d\tilde{x}^\ell}{dt}$$

$$= -2\frac{\partial V}{\partial x^\ell}\frac{d\tilde{x}^\ell}{d\tilde{s}}\frac{d\tilde{s}}{dt}$$

$$= -4\frac{\partial V}{\partial x^\ell}\cdot(h-V)\frac{d\tilde{x}^\ell}{d\tilde{s}}\ . \qquad (3)$$

(5) Substitute in (2) for $d\tilde{s}/dt$ from (3), and for $d^2\tilde{s}/dt^2$ from (4).

(6) Hence

$$\tilde{g}_{kj}\frac{d^2\tilde{x}^j}{d\tilde{s}^2} + \tilde{\Gamma}_{ij,k}\frac{d\tilde{x}^i}{d\tilde{s}}\frac{d\tilde{x}^j}{d\tilde{s}} = 0,$$

which implies the assertion of the proposition.

15.4. Proposition 15.3 has the following converse:

Let $\tilde{x}(\tilde{s})$ be any geodesic for the Jacobi metric, and define the parameter t, the "time variable", by

$$\frac{dt}{d\tilde{s}} = \frac{1}{2[h-V(\tilde{x}(\tilde{s}))]} .$$

Then the function $t \to x(t)$ defined by

$$x(t) = \tilde{x}(\tilde{s}) \qquad \text{(Def.)}$$

is a solution to Lagrange's equations.

Proof: Reverse the proof of Proposition 15.3.

15.5. Since a Lagrange curve can, by Proposition 15.3, be regarded as a geodesic (with respect to the appropriate Jacobi metric), the results of §14 about geodesics apply to Lagrange curves. Thus the following propositions are obtained:

For any point $\overset{o}{x}$ of a Lagrangian manifold, there is a neighbourhood U such that each point $\overset{1}{x}$ of U is connected with $\overset{o}{x}$ by exactly one Lagrange curve $t \to x(t)$, with $\overset{o}{x}$ and $\overset{1}{x}$ as its endpoints.

15.6. Using the notation of the preceding propositions, let

$$x(\overset{o}{t}) = \overset{o}{x}, \qquad x(\overset{1}{t}) = \overset{1}{x} ,$$

and $\quad y(t), \quad \overset{o}{t} \le t \le \overset{1}{t}$, be any curve with

$$y(\overset{o}{t}) = \overset{o}{x}, \qquad y(\overset{o}{t}) = \overset{1}{x} .$$

Then, for all $\overset{1}{x}$ sufficiently close to $\overset{o}{x}$,

$$\int_{\overset{o}{t}}^{\overset{1}{t}} \sqrt{\tfrac{1}{2}(h-V(y))g_{ij}(y)\,\frac{dy^i}{dt}\,\frac{dy^j}{dt}}\;dt$$

$$\ge \int_{\overset{o}{t}}^{\overset{1}{t}} \sqrt{\tfrac{1}{2}(h-V(x))g_{ij}(x)\,\frac{dx^i}{dt}\,\frac{dx^j}{dt}}\;dt \quad .$$

Equality holds if, and only if, the ranges of the arcs $t \to x(t)$ and $t \to y(t)$ are identical. The right-hand side of the inequality is equal to

$$\int_{t_0}^{t_1} G(x, \frac{dx}{dt})dt$$

where

$$G(x,\dot{x}) = \tfrac{1}{2}\, \dot{x}^* g(x)\dot{x}. \qquad \text{(Def.)}$$

It is the "kinetic energy" along the geodesic arc $x(t)$. The time integral of the kinetic energy is called action. Hence the following corollary to Proposition 15.6, called the Principle of Least Action:

15.7. Suppose that the curve $y(t)$ satisfies the additional condition that

$$G(y, \frac{dy}{dt}) + V(y) = h.$$

Then,

$$\int_{t_0}^{t_1} G(y, \frac{dy}{dt})dt \geq \int_{t_0}^{t_1} G(x, \frac{dx}{dt})dt$$

and

$$\int_{t_0}^{t_1} L(y, \frac{dy}{dt})dt \geq \int_{t_0}^{t_1} L(x, \frac{dx}{dt})dt$$

where

$$L = G - V$$

is the Lagrange function of the system.

Proof of the last inequality: Note that

$$L = G - V = 2G - (G + V) = 2G - h,$$

and apply the preceding inequality.

References

Laugwitz (1965), Milnor (1969).

§16. Spaces With Intrinsic Metrics

16.1. Propositions 14.11 and 14.19 have lead us to the following situation:

A connected Hausdorff space with a countable basis (a Riemann manifold) is given, and for each of its points there exists a neighbourhood such that any two of its points have a distance (the length of the geodesic arc connecting both those points). Moreover, if both these points are sufficiently close to each other, any other arc connecting the points has a length of arc larger than the distance of the points.

Such a distance function ρ defined in a given neighbourhood V satisfies the three postulates for a metric space:

(i) For any points x and y of V ,

$$\rho(x,y) = 0 \text{ if, and only if, } x = y.$$

(ii) For any $x,y \in V$,

$$\rho(x,y) = \rho(y,x).$$

(iii) For any points x,y, and z of V,

$$\rho(x,z) \leq \rho(x,y) + \rho(y,z). \quad (\text{"Triangle inequality."})$$

The metric space will be denoted by the symbol (V,ρ).

16.2. A metric space (V,ρ) is endowed with a topology in a way which is a straight-forward generalization of the procedure indicated in §1.4:

Let ε be any positive real, and x be any point of V. The set

$$U(x,\varepsilon) \underset{\text{Def.}}{=} \{y \in V | \rho(y,x) < \varepsilon\}$$

is called the ε-neighbourhood, or a spherical neighbourhood, of the point x. The terms "open ε-ball" and "open ball" about x will also be used. A subset

S of V is said to be open (with respect to the given distance function ρ) if for each point x of S there is a spherical neighbourhood of x which is a subset of S. In particular, the spherical neighbourhoods themselves are open, because of the triangle inequality.

16.3. Suppose that for each point x of a topological space (X,Γ) a neighbourhood V_x is given in which a distance function ρ_x is defined so that (V_x,ρ_x) is a metric space. Suppose further that, if any two of such "metric neighbourhoods", (V_x,ρ_x) and (V_y,ρ_y) intersect, their distance functions ρ_x and ρ_y are equal in the intersection, $V_x \cap V_y$. For this we shall say: the local distance functions are mutually compatible.

The family of all metric neighbourhoods generates (i.e., is a basis of) a topology for the space, just as the open balls of a metric space form a basis of a topology (§16.2): A subset S of X is called open if, for any point x of S, there is a spherical neighbourhood of x lying in $S \cap V_x$. If this new topology with the basis of metric neighbourhoods $\{(V_x,\rho_x)|x \in X\}$ is identical with the original topology Γ of (X,Γ), we shall say the family of metric neighbourhoods $\{(V_x,\rho_x)|x \in X\}$ is topologically compatible with the topology Γ.

Thus, in particular, the family of metric neighbourhoods in a Riemann manifold, discussed in §14.11, is topologically compatible with the topology of the manifold.

16.4. A Hausdorff (X,Γ) space will be said to be a locally metric space if

(i) for each of its points there exists a metric neighbourhood,

(ii) all the distance functions of these metric neighbourhoods are mutually compatible, and

(iii) the family of the metric neighbourhoods is topologically compatible with the topology Γ.

Any metric space is a locally metric space.

16.5. An arc in a given Hausdorff space (X,Γ) is defined as a continuous mapping from the closed unit interval [0,1] into the space X. As a continuous image of a connected space, the range of an arc is a connected subset of X. A subset of X is said to be arcwise connected if any two points of the subset are connected by an arc whose range lies entirely in that subset. The space is said to be locally arcwise connected if for each of its points x and each open neighbourhood U of x, there is an arcwise connected open neighbourhood V of x such that $V \subseteq U$. An indirect argument shows

An arcwise connected Hausdorff space is connected.

16.6. A connected Hausdorff space which is locally arcwise connected is arcwise connected.

Proof. Let x be any point of the space X, and A(x) be the set of all points of X which can be connected with x by an arc. The set A(x) is both open and closed. Since X is connected, its only subsets which are both open and closed are the empty set and X itself. Because $x \in A(x)$, A(x) is not empty. Hence, A(x) = X.

Corollary: Lagrange and Riemann manifolds are arcwise connected.

16.7. For our further discussion the concept of compactness will be needed. Let Δ be a family of open subsets of a topological space (X,Γ) whose union is equal to the space X. Then Δ is called an open covering of X. A subfamily of Δ which is also an open covering of X is said to be an (open) subcovering. The space (X,Γ) is said to be compact if every open covering of X contains a finite subcovering. A subset Y of (X,Γ) is called compact if Y, considered as a subspace endowed with the relative topology, is compact.

Example. Let (X,Γ) be a closed bounded interval of the real line with the "usual" topology introduced in §1.4. Then the Heine-Borel Theorem states that (X,Γ) is compact. Any closed and bounded subset of the real line is compact.

Let the mapping f from (X_1,Γ_2) onto (X_1,Γ_2) be continuous. Then, if (X_1,Γ_1) is compact, (X_2,Γ_2) also is compact.

This proposition is a straight-forward consequence of the fact that by §1.6, the pre-image $f^{-1}(G_2)$ of any open subset G_2 of (X_2,Γ_2) is an open subset of (X_1,Γ_2).

Corollary: The range of an arc is a compact set.

16.8. The length of an arc in an arcwise connected, locally metric Hansdorff space (X,Γ) will now be defined.

Let $x(t)$, $0 \leq t \leq 1$, be an arc. Assign, to each point of its range, a metric neighbourhood, and select from this open covering of the range a finite sub-covering. Choose a finite strictly increasing sequence $\{t_i\}_{i=0}^{n}$ of points of the unit interval such that $t_0 = 0$, $t_n = 1$, and all local distances $\rho(x(t_{i-1}), x(t_i))$, $i=1,2,\cdots,n$, are defined. Call the (finite or infinite) supremum of all the sums $\sum_{i=1}^{n} \rho(x(t_{i-1}), x(t_i))$ obtainable in this fashion the length of the arc $t \to x(t)$. If this length is finite, the arc is said to be rectifiable.

16.9. Two arcs $t \to x(t)$ and $t' \to x'(t')$ are said to be equivalent if there is a continuous one-to-one mapping $t' = \phi(t)$, $0 \leq t \leq 1$, such that $\phi(0) = 0$, $\phi(1) = 1$, and $x'(\phi(t)) = x(t)$ for all $t \in [0,1]$. This relation between arcs is an equivalence relation. If one of two equivalent arcs is rectifiable, so is the other one, and both have the same length. It may, therefore, also be assigned to the common range of both the arcs as its length. Again, if the length is finite, the range is said to be rectifiable. In this case, there is a one-to-one correspondence between the points $x(t)$ of the range $\{x(t) \in X | t \in [0,1]\}$ and the lengths $s(t)$ of the arcs $\tau \to x(\tau)$, $0 \leq \tau \leq t$. Thus the length of arc s can be used for a parametrization $s \to x$ of the point set $\{x(t) \in X | t \in [0,1]\}$. A rectifiable arc with this parametrization will be called a normal arc. The given arc $t \to x(t)$ and the corresponding normal arc are equivalent if, and only if, the function $t \to x(t)$ is nowhere constant; that is, for each $t \in [0,1]$ there exists a neighbourhood (in $[0,1]$) in which the function $x(t)$ is one-to-one.

16.10. A locally metric space is said to be finitely arcwise connected if any two of its points can be connected by a rectifiable arc.

An arcwise connected locally metric Hausdorff space (X,Γ) such that its local metric neighbourhoods are finitely arcwise connected, is a finitely arcwise connected metric space.

Proof. Let $x(t)$, $0 \leq t \leq 1$, be an arc connecting any two given points a and b of X. The range of this arc can, as a compact subset of X, be covered by finitely many metric neighbourhoods. Choose points $x(t_i)$, $i=0,\cdots,n$, such that $t_0 = 0$, $t_n = 1$, and, for each $i \geq 1$, $x(t_{i-1})$ and $x(t_i)$ belong to the same metric neighbourhood. Let $y_i(\tau)$, $0 \leq \tau \leq 1$, $i=1,2,\cdots,n$, be rectifiable arcs connecting the points $x(t_{i-1})$ and $x(t_i)$. Thus a and b are connected by a rectifiable arc; hence, (X,Γ) is finitely arcwise connected.

Now a distance function ρ in X can be defined: let $\rho(a,b)$ be the infimum of the lengths of all rectifiable arcs connecting a and b. The three conditions of §16.1 for a a distance function are satisfied by ρ.

16.11. Let (X,ρ) be a finitely arcwise connected metric space, and define the function $\rho_i : X \times X \to R$ by

$$\rho_i(x,y) = \inf\{\text{lengths of all the rectifiable arcs connecting } x \text{ and } y\}.$$

The function ρ_i is called the intrinsic distance function of the metric space (X,ρ), and it is a distance function in the sense of §16.1; that is, for any points x,y,z of X:

(i) $\rho_i(x,x) = 0;$

(ii) $\rho_i(x,y) = \rho_i(y,x);$

(iii) $\rho_i(x,z) \leq \rho_i(x,y) + \rho_i(y,z).$

Moreover,

$$\rho(x,y) \leq \rho_i(x,y)$$

and, consequently,

$$\{y \in X | \rho_i(x,y) < \varepsilon\} \subseteq \{y \in X | \rho(x,y) < \varepsilon\}.$$

Thus, every ρ-open subset of X (i.e., every subset of X which is open in the topology generated by the distance function ρ) is also ρ_i-open; the reverse may, but need not, be true.

In the following, it will be assumed that the distance functions of the metric spaces considered are intrinsic distance functions. Such a space will be called a space with an intrinsic metric. By its very definition it is assumed to be a finitely arcwise connected metric space.

16.12. Suppose that the local distance functions mentioned in Proposition 16.10 are intrinsic. Then the global distance function constructed, in §16.10, for the whole space (X,Γ) is also intrinsic.

Further, the local distance function of any of the local metric space is identical with the global distance function restricted to that metric neighbourhood.

The topology generated by the global distance function is identical with the given topology Γ of the Hausdorff space (X,Γ).

16.13. A bounded closed subset of E^m is compact, and, conversely, any of its compact subset is closed and bounded.

This fact motivates the following definitions:

Let S be any non-empty subset of a given metric space (X,ρ). Then the diameter of the set S is defined as

$$\sup \{\rho(x,y) \mid x \in S,\ y \in S\}.$$

A set whose diameter is finite is said to be bounded. If every bounded closed subset of X is compact, and every compact subset is closed and bounded, then the metric space (X,ρ) is said to be finitely compact. In particular, any compact metric space is finitely compact.

Suppose that the distance function of the metric space (X,ρ) is intrinsic, and that (X,ρ) is finitely compact. Then for any two points x and y of X there is a rectifiable arc connecting x and y whose length is equal to $\rho(x,y)$. The length of any other arc connecting both those points is $> \rho(x,y)$.

For the proof see the literature.

16.14. Riemann manifolds are finitely compact spaces with intrinsic distance functions. Hence Proposition 16.13 applies by which any two points of a Riemann manifold can be connected by an arc of least length. If the points are sufficiently close to each other, the connecting shortest arc is uniquely determined, by Proposition 14.11.

16.15. Thus the theory of dynamical systems with Lagrange functions of the form

$$L(x,\dot{x}) = \tfrac{1}{2}\, \dot{x}^* G(t) \dot{x} - V(x)$$

has been imbedded first, in the geometry of Riemann manifolds (§15), and this , in turn, has been imbedded in the "intrinsic geometry" of metric spaces, as sketched in this section. Hence the results of intrinsic geometry apply to dynamical systems of the above kind.

Thus, for example, the existence of closed geodesics can be proven for a certain class of dynamical systems; that is, it can be shown by methods of intrinsic geometry that for these dynamical systems periodic notions exist, and a lower bound of their number is obtained from the topological structure of the Lagrange manifold. (It equals the number of homotopy classes on the manifold).

References

For the intrinsic geometry of metric spaces see Rinow (1961). A proof for the existence of geodesic curves is also found in Kolmogorov-Fomin (1957). Lee (1970) proved and generalized a theorem by Whittaker (1961) about the existence of periodic metrics by those methods. The complicated analytic condition stated by Whittaker turned out to have a simple geometric meaning. A less rigorous discussion is found in Birkhoff (1927). For other methods of finding periodic motions see Birkhoff (1927), Nemytskii-Stepanov (1960), Siegel (1956), Siegel-Moser (1971).

§17. A Generalization of Jacobi's Interpretation

17.1. The geometric interpretation of a Lagrange curve as a geodesic can be generalized to the case that the Lagrange function L is of the form

$$L(x,\dot{x}) = \frac{1}{2}\dot{x}^* g(x,\dot{x})\dot{x} - V(x)$$

where $g(x,\dot{x}) = \{g_{ij}(x,\dot{x})\}_{i,j=1}^{m}$ is a positive definite matrix of the order m which is homogeneous in $\dot{x}$ of the order 0; that is,

$$g(x,\lambda\dot{x}) = g(x,\dot{x})$$

for any positive real λ.

For the following discussion note that, by Euler's identity,

$$\frac{\partial g_{ij}(x,\dot{x})}{\partial \dot{x}^k}\,\dot{x}^k = 0\ .$$

First a generalization of the notion of a Riemann manifold will be given.

17.2. Let M be a connected analytic manifold, and D be a coordinate neighbourhood on M. For a given system of coordinates, let $g(x,\dot{x})$ be a positive definite symmetric matrix of the order m such that

$$g(x,\lambda\dot{x}) = g(x,\dot{x})$$

for every positive real λ. A straight-forward computation shows that $g(x,\dot{x})$ is a representation of a covariant tensor of the second order - denoted, for convenience, by the same symbol. It follows that the sum

$$g_{ij}(x,\dot{x})\,\frac{dx^i}{dt}\,\frac{dx^j}{dt}$$

is, for any given curve $t \to x(t)$, invariant under (admissible) transformations of local coordinates. In generalization of the notion of the length of arc in Riemannian geometry, the length of arc, s, of the curve x(t) may therefore be defined by

$$\frac{ds}{dt} = \sqrt{g_{ij}(x,\dot{x})\,\frac{dx^i}{dt}\,\frac{dx^j}{dt}}\ ,$$

since the right-hand side of this equation has a meaning which is independent of the choice of the system of coordinates. Because of the assumed homogeneity of g in $\dot{x}$ the length of arc is also invariant under parameter transformations: hence, it may be ascribed to the range of the arc (i.e. the point-set arc).

Suppose that for each coordinate neighbourhood such a "metric tensor" has been defined, and that for any two intersecting neighbourhoods these metric tensors are identical in the intersection. Then we shall call the manifold M, endowed with the metric $g(x,\dot{x})$, a <u>Finsler manifold</u> (M,g).

<u>17.3.</u> A <u>geodesic</u> of such a manifold is defined, as in Riemannian geometry (§14.8), as a curve of vanishing first variation.

<u>The equation of a geodesic in a Finsler manifold (M,g) reads</u>

$$\frac{d^2x^k}{ds^2} + \Gamma^k_{ij} \frac{dx^i}{ds}\frac{dx^j}{ds} = 0 , \qquad k=1,\cdots,n,$$

just as in Riemann geometry. The (generalized) Christoffel symbols of the first and second kinds, $\Gamma_{ij,k}$ and Γ^k_{ij} , are defined by the same formulas as in Riemannian geometry (§14.8), and s again denotes the length of arc.

The proofs of this and the following propositions are similar to the proofs of the corresponding theorems of Riemannian geometry.

<u>17.4.</u> <u>The geodesic arc between two points of a Finsler manifold which are sufficiently close to each other is shorter than any other arc connecting those points.</u>

<u>17.5.</u> The Hamilton function H corresponding to the Lagrange function L of §17.1 is given by

$$L + H = \dot{x}^* y$$

where

$$y \underset{\text{Def.}}{=} L_{\dot{x}} .$$

Hence, by Euler's identity (§17.1),

$$L + H = \dot{x}^* g\dot{x}.$$

Now, H is a first integral of the Hamilton equations; therefore,

$$Q(x,\dot{x}) \underset{\text{Def.}}{=} H(x,y)$$

is a first integral of the corresponding Lagrange equations. But

$$\begin{aligned} Q(x,\dot{x}) &= \dot{x}^* g\dot{x} - L \\ &= \frac{1}{2}\dot{x}^* g\dot{x} + V. \end{aligned}$$

Hence, the function

$$\frac{1}{2}\dot{x}^* g(x,\dot{x})\dot{x} + V(x)$$

is a first integral of the Lagrange equations.

17.6. There are the following generalizations of Propositions 15.1 and 15.4:

Let h be the value taken by the "energy integral"

$$Q(x,\dot{x}) \underset{\text{Def.}}{=} \frac{1}{2}\dot{x}^* g(x,\dot{x})\dot{x} + V(x)$$

on the Lagrange curve $t \to x(t)$; further define, on the set $\{x \in M | V(x) < h\}$, the Finsler metric $\tilde{g}$ by

$$\tilde{g}(x,\dot{x}) = 2(h - V(x))g(x,\dot{x}),$$

and let

$$y(t) = x(t) + \varepsilon\xi(t),$$

where ε is a real parameter, be any curve such that

$$\begin{cases} y(t^0) = x(t^0), \\ y(t^1) = x(t^1) \end{cases}$$

and

$$Q(y, \frac{dy}{dt}) = h + Q(\varepsilon^2) \text{ as } \varepsilon \to 0.$$

(t^0 and $t^1 > t^0$ are any two admissible values of the parameter t.) Then,

$$\left[\frac{\partial}{\partial \varepsilon} \int_{t^0}^{t^1} \sqrt{\frac{dy^*}{dt} \; \tilde{g}(y, \frac{dy}{dt}) \frac{dy}{dt}} \, dt \right]_{\varepsilon=0} = 0.$$

17.7. Assume the hypotheses and notation of Proposition 17.6 , let s denote the length of arc of the Lagrange curve $t \to x(t)$ in the appropriate Finsler metric $\tilde{g}$, and define $\tilde{x}$ by $\tilde{x}(s) = x(t)$. Then,

$$\frac{d^2x^k}{ds^2} + \tilde{\Gamma}^k_{ij} \frac{dx^i}{ds} \frac{dx^j}{ds} = 0,$$

where the $\tilde{\Gamma}$'s are the Christoffel symbols of the second kind for the metric $\tilde{g}$.

17.8. In a Finsler manifold an intrinsic metric can be defined in the same way as in the special case of a Riemann manifold, and the resulting topology is identical with the original topology of the manifold. Thus the theory of dynamical systems with Lagrange functions of the kind introduced in §17.1 can be interpreted in terms of the intrinsic geometry of metric spaces.

References

For Finsler geometry see, for example, Rinow (1961), Laugwitz (1965), Rund (1959).

§18. Concluding Remarks

18.1. Another generalization of the notion of a dynamical system is obtained in the following way.

Let G be a given region of the n-dimensional Euclidean space E^n, and

$$x = \phi(\overset{o}{x}, t), \quad \overset{o}{x} \in G, \; -\infty < t < +\infty ,$$

be the general solution of the differential equation

$$\frac{dx}{dt} = f(x), \quad x \in G,$$

with

$$\phi(\overset{\circ}{x},0) = \overset{\circ}{x},$$

and suppose that

$$\phi(\overset{\circ}{x},t) \in G$$

for all admitted $\overset{\circ}{x}$ and t. Thus for each value of t there corresponds a mapping

$$\phi^t \underset{\text{Def.}}{=} \phi(\cdot,t)$$

of G into itself, and by §10.5 all these mappings form a one-parametric and, therefore, commutative "*transformation group*": the group operation is the composition of two mappings,

$$\phi^r\phi^s = \phi^{t+s},$$

which implies the associative and commutative laws, ϕ^0 is the identity mapping, and ϕ^{-t} is the inverse of ϕ^t. Moreover, $\phi^t\overset{\circ}{x}$ is a continuous function of $(\overset{\circ}{x},t)$.

Now let X be a connected metric space or, more generally, a connected Hausdorff space, and $\{\phi^t : X \to X \mid -\infty < t < \infty\}$ be a group of mappings having the properties just described.

The theory of such groups, called *topological dynamics*, yields qualitative results of a topological nature (for example, about stability) which apply immediately to dynamical systems whose Lagrange equations satisfy the above conditions. Such a group may itself be called a *dynamical system* in a generalized sense.

18.2. The topological approach can also be applied to the following kind of dynamical systems. Suppose that the solutions $\phi^t\overset{\circ}{x}$ of the equation

$$\frac{dx}{dt} = f(x), \quad x \in G \subseteq E^n$$

are defined and unique only in open intervals $T_{\overset{\circ}{x}}$ which may, but need not, be the whole time axis. Then, for any such function $f : G \to E^n$, a positive real function α can be constructed such that, for any starting point $\overset{\circ}{x}$ in G, the equation

$$\frac{dy}{dt} = \frac{f(y)}{\alpha(y)}, \qquad y \in G \subseteq E^n$$

has a unique solution, $\psi^t x$, defined for all real t. The function α is defined in such a way that, if the moving point y(t) approaches the boundary of G, the values of α increase so rapidly, the motion becomes so slow, that the moving point never reaches the boundary. The point sets $\{\phi^t \overset{\circ}{x} | t \in T_{\overset{\circ}{x}}\}$ and $\{\psi^t \overset{\circ}{x} | t \in T\}$ are identical: for, a suitable division reduces either system of n scalar differential equations to the same system of (n-1) equations, eliminating the time variable t and the function α simultaneously, and this proves the identity of those point sets. The topological results concerning the ranges of the solution curves $t \to y(t)$ can immediately be transferred to the ranges of the solution curves $t \to x(t)$ of the original equation for dx/dt.

18.3. Next, we turn to non-autonomous differential equations of the form

$$\frac{dx}{dt} = f(x,t).$$

By the device of §10.5, such an equation is transformed into the equivalent autonomous system

$$\begin{cases} \dfrac{dx}{ds} = f(x,t), \\ \dfrac{dt}{ds} = 1, \end{cases}$$

and now the procedure of the preceding articles can be applied. Thus the transformation groups are here of the form $\{\chi^s : X \times T \to X \times T | -\infty < s < \infty\}$. The distance function of $X \times T$ can be defined, for example, as the sum of the distance functions of X and T.

18.4. This approach could also be combined with the geometric interpretation discussed earlier in this chapter:

Let X and Y be two non-empty sets, $(X \times Y, \rho)$ be a metric space,

$\chi^t: X \times Y \to X \times Y$, for $-\infty < t < \infty$, be a one-parametric transformation group, I be an invariant subset of $X \times Y$ under this group so that $\chi^t I = I$ for all t, and suppose there is an intrinsic metric σ for the subspace X of the following kind: the topology of X determined by σ is identical with the topology determined by the restriction of ρ to X, and with respect to this metric σ any "orbit" $t \to \phi^t(\mathring{x},\mathring{y})$, where $(\mathring{x},\mathring{y})$ is an arbitrary point of I and $\phi^t(\mathring{x},\mathring{y})$ is the projection of

$$\chi^t(\mathring{x},\mathring{y}) = (\phi^t(\mathring{x},\mathring{y}),\ \psi^t(\mathring{x},\mathring{y}))$$

into X, is a "geodesic", i.e. a curve of shortest local arcs. For any two real numbers α and $\beta > \alpha$ sufficient close to each other, the arc $\phi^t(\mathring{x},\mathring{y})$, $\alpha \leq t \leq \beta$, has the length $\sigma(\phi^\alpha(\mathring{x},\mathring{y}),\ \phi^\beta(\mathring{x},\mathring{y}))$.

These assumptions yield a notion of a dynamical system which is intermediate between Jacobi's interpretation of a conservative Lagrangian system (§15) and the concept of a dynamical system as a transformation group (§18.1). It seems that this approach has not yet been pursued.

Notes and References

§18.1. Standard references are Nemytskii-Stepanov (1960) and Gottschalk-Hedlund (1955). In the latter book often general topological transformation groups are assumed, acting on general topological spaces. (For topological groups see Pontrjagin (1958)). In Nemytskii-Stepanov's book only metric space and one-parametric transformation groups are considered. This relative specialization in the hypotheses yields more significant, and still very general, results. A bibliography of topological mechanics has been given by Gottschalk (1966). Later results are found in Auslander-Gottschalk (1968).

§18.2. For the transformation of the differential equation indicated in the text, due to Vinograd see, for example, Nemytskii-Stepanov (1960), pp. 19-21.

EXERCISES

§1

1. Prove the assertions (i) and (ii) of §1.4.
2. Is the intersection of any (non-empty) family of open sets in E^n open? (§1.4.)
3. Is the topological space E^n Hausdorff? Is it connected? (§1.5.)
4. Prove that the homeomorphism of topological spaces is an equivalence relation. (§1.6.)
5. Prove that the homeomorphic image of an open set is an open set. (§1.6.)
6. Prove both the assertions of §1.7.
7. Prove that the local coordinate transformations of differentiability class C^∞ form a group. (§1.10.)

§2

8. Prove the invariance of the centre of masses under the transformations of coordinates mentioned in §2.4 (iii).
9. Let $t \to (\underline{x}_1, \underline{x}_2, \cdots, \underline{x}_p)$ be any solution of the equations of motion of a free dynamical system of particles (§2.4.) Prove that

$$\frac{d}{d\underline{t}} \sum_{j=1}^{p} \mu_j \frac{d\underline{x}_i}{d\underline{t}} = 0 .$$

10. Let, in Exercise 9, $\underline{x}_1, \cdots, \underline{x}_p$ be 3-vectors, and denote vector products (cross products) by "×". Prove that

$$\frac{d}{d\underline{t}} \sum_{j=1}^{p} \mu_j \frac{d\underline{x}_i}{d\underline{t}} \times \underline{x}_j = 0 .$$

11. Derive, for a free dynamical system of two particles, the equation of relative motion, i.e. the differential equation for $\underline{y} \underset{\text{Def.}}{=} (\underline{x}_2 - \underline{x}_1)$ (§2.6.), and show that it can be interpreted as an equation of motion of a one-particle system (§2.5.)

12. Let $t \rightarrow \underline{y}(t)$ be any solution of the differential equation derived in Exercise 11. Show that all the vectors $\underline{y}(t)$ lie in some fixed plane.

13. Give all the details for the proof of the Existence and Uniqueness Theorem of §2.9.

14. In the proof outlined in §2.9 of the Existence and Uniqueness Theorem it has (as always) been assumed that the given function g has derivatives of all orders. Is that assumption indispensable? Is the assumption that g is continuous sufficient?

15. Consider the one-particle system of Exercise 11. Assume that, at some instant, the direction of the motion (i.e. the direction of the vector $d\underline{y}/d\underline{t}$) is pointed towards the origin. Prove that then it is so always.

§3

16. Derive Lagrange's function and Lagrange's equations of motion of a one-particle system (cf. Exercise 11) in E^3 for polar coordinates.

17-19. Prove Propositions 3.4, 3.5, and 3.7.

20. Prove that the relations "is equivalent to" introduced in §3.9 and §3.13 are equivalence relations.

21. Prove that $\underline{L}_{\dot{\underline{x}}*\dot{\underline{x}}}$ is non-singular if, and only if, $\overline{\underline{L}}_{\dot{\overline{\underline{x}}}*\dot{\overline{\underline{x}}}}$ is non-singular. (§3.19.)

22. Let

$$\underline{L}(\underline{x},\dot{\underline{x}}) = \tfrac{1}{2} \sum_{i=1}^{m} gi(x^i)\, (\dot{x}^i)^2 + \sum_{i=1}^{m} V_i(x^i) .$$

Integrate the corresponding Lagrange equations. (§3.6.)

§4

23. Take the function $\underline{y}^*\dot{\underline{x}} - \underline{H}(\underline{x},\underline{y},t)$ as a Lagrange function of the *independent* variables $\underline{x},\underline{y},\dot{\underline{x}},\dot{\underline{y}},t$ and derive the corresponding Lagrange equation. (§4.2.)

24. Prove Proposition 4.6.

25. Derive the Hamilton function and the Hamilton equations corresponding to the Lagrange equations obtained in Exercise 16.

§5

26. Prove Proposition 5.8.

27. Verify that the transformation of §5.14 is canonical with multiplier μ.

§6

28. Prove Corollary 6.3.

§7

29. Prove that the differential form $Z^*(z)dz$ is an exact differential if, and only if, the matrix $Z_{z^*}(z)$ is symmetric. (§7.2.)

30. Give the full proof of Proposition 7.2.

31. Complete the proof of Lemma 7.4.

32. Prove that the transformation of §7.6 is canonical.

33-34. Prove Propositions 7.7 and 7.8.

§8

35. Prove Proposition 8.3.

§9

36. Work out the details of the proof in §9.2.

§10.

37. Is a derivative of a first integral a first integral? (§10.3.)

38. Integrate the equation of motion of a freely falling particle, and determine all its first integrals by an application of Proposition 10.4.

39. Prove Proposition 10.8.

40. Prove Proposition 10.3 by means of Proposition 10.8.

41. Prove that (n+1) first integrals are functionally dependent, by using Liouville's equation. (§10.8.)

42. Reduce the equation of motion of a free dynamical system of particles by the "integral of momentum" introduced in Exercise 9. (§10.9.)

43. Prove Proposition 10.11 by Liouville's equation 10.8.

44. Prove Proposition 10.13.

45. Formulate and prove the "analogous proposition for Lagrange equations" mentioned in §10.17.

46. Explain the "slight change of the meaning of the symbol U" mentioned in the second paragraph of §10.19.

47. Prove Proposition 10.19.

§11

48. Supply the details omitted in the proof of Proposition 11.5.

49. Is the functional $\int_0^1 t^\alpha \frac{dx}{dt}\, dt$, with $x(t) \in E^1$, $\alpha = \text{const} \geq 0$, invariant under the family of transformations

$$\begin{cases} \bar{x} = x + a\varepsilon, & (a = \text{const.}) \\ \bar{t} = t + \varepsilon \; ? \end{cases}$$

(§11.5.)

50. Complete the proof outlined at the end of §11.8.

§12

51. Prove Proposition 12.6 first for a time-independent, then for a time dependent "Liapunov function" F.

52. Prove that the functions λ' and λ'' of Proposition 12.10 are continuous and, therefore, integrable functions of t.

53. Prove the equation for $|z(t)|$ given at the beginning of the proof in §12.11.

54-56. Complete the proofs of Propositions 12.11-12.13.

§13

57. Give the details of the proof in §13.3.

§14

58. Prove that the integral yielding the length of arc is invariant under parameter transformations. (§14.4.)

59. Determine the Christoffel symbols of both kinds for Cartesian and for polar coordinates in the Euclidean plane. (14.8.)

60. Are the Christoffel symbols components of a (representative of a) tensor? (Cf. Exercise 59).

§15

61-62. A free particle upon which no force is acting moves on a cylinder (on a sphere) in E^3. Give all types of possible motions. (§15.3.)

§16

63. Prove that the range of an arc is connected. (§16.5.)

64. Prove that the equivalence of arcs (§16.9) is an equivalence relation.

§17

65. Prove the last identity in §17.1.

66. Derive the equation of a geodesic in a Finsler manifold. (§17.3.)

REFERENCES

Auslander, J., and W. H. Gottschalk

1968 Topological Dynamics. An International Symposium.
W. A. Benjamin, New York-Amsterdam

Bellman, R.

1953 Stability Theory of Differential Equations.
McGraw-Hill. Reprint by Dover Publications, New York

Birkhoff, G. D.

1927 Dynamical Systems.
Amer. Math. Soc. Coll. Publ. Vol 9

Bliss, G. A.

1962 Calculus of Variations.
The Mathematical Association of America

Bourbaki, N.

1966 General Topology.
Hermann, Paris

1968 Theory of Sets.
Hermann, Paris

Bunge, M.

1967 Foundations of Physics.
Springer-Verlag, New York

Carathéodory, C.

1935 Variationsrechnung und partielle Differentialgleichungen erster Ordnung.
Teubner, Leipzig-Berlin

Charlier, K.

1902 Die Mechanik des Himmels.

Von Veit, Leipzig

Courant, R., and D. Hilbert

1953 Methods of Mathematical Physics, Vol. I.

Interscience Publ., New York-London

1962 Methods of Mathematical Physics, Vol. II.

Interscience Publ., New York-London

Fraenkel, A.

1966 Abstract Set Theory.

North Holland Publ. Co.

Frank, P.

1969 Foundations of Physics. (6th ed.)

U. Chicago Press

Gelfand, I. M., and S. V. Fomin

1963 Calculus of Variations.

Prentice-Hall, Englewood Cliffs, N. J.

Gottschalk, W. H.

1966 Bibliography for Topological Dynamics.

Wesleyan University, Middletown, Connecticut

Gottschalk, W. H., and G. A. Hedlund

1955 Topological Dynamics.

Amer. Math. Soc. Coll. Publ. Vol 36

Hahn, W.

1963 Theory and Applications of Liapunov's Direct Method.

Prentice-Hall, Englewood Cliffs, N. J.

1967 Stability of Motion.

Springer-Verlag, New York

Hausdorff, F.

1914 Grundzüge der Mengenlehre.

Berlin. Reprint by Chelsea Publ. Co., New York

Heller, K. D.

1964 Ernst Mach.

Springer-Verlag, Berlin-Göttingen-Heidelberg

Herstein, I. N.

1964 Topics in Algebra.

Blaisdall Publ. Co., Waltham, Mass.

Hurewicz, W., and H. Wallman

1948 Dimension Theory.

Princeton U. P.

Jammer, M.

1969 Concepts of Space.

Harvard U. P.

Joos, G.

1959 Theoretical Physics.

Blackie, London

Kamke, E.

1928 Mengenlehre.

W. de Gruyter, Berlin

Kant, J.

1786 Metaphysische Anfangsgründe der Naturwissenschaft.

Kelley, J. L.

1955 General Topology.

Van Nostrand, Princeton

Khinchin, A. J.

1949 Mathematical Foundations of Statistical Mechanics.

Dover, New York

Kolmogoroff, A. N., and S. V. Fomin

1957 Elements of the Theory of Functions and Functional Analysis, Vol. I.
Graylock Press, Rochester, N. Y.

Kurth, R.

1952 Das Anfangswertproblem der Stellardynamik.
Zeitschrift f. Astrophysik 30, pp. 213-229

1955 Verallgemeinerung eines elementaren Satzes von Laplace.
Elem. d. Mathematik 10, pp. 127-128

1957a Kants Lehre von Raum und Zeit.
Philosophia Naturalis 4, pp. 266-296

1957b Introduction to the Mechanics of Stellar Systems.
Pergamon Press, London

1958 On Lagrange's triangular solution of the problem of three bodies.
Archiv d. Mathematik 8, pp. 381-392

1959 Introduction to the Mechanics of the Solar System.
Pergamon Press, London

1960a Axiomatics of Classical Statistical Mechanics.
Pergamon Press, London

1960b On the Foundations of Celestial Mechanics.
Astron. Nachr. 285, pp. 229-232

1962 Masse als Menge der Materie.
Philosophia Naturalis 7, pp. 356-364.

1964a Uber Zeit und Zeitmessung.
Philosophia Naturalis 8, pp. 65-90

1964b Uber das Newtonsche Bewegungsgesctz.
Philosophia Naturalis 8, pp. 459-461

1966a On a Theorem by Laplace.
Comm. Phys.-Math. Soc. Sc. Fenn. 33 Nr. 3

1966b On a Theorem by Lagrange.

Proc. Kon.-Ned. Ak. Wet. Amsterdam, 1969, pp. 240-250

1971a Set-Theoretical Aspects of the Restricted Problem of Three Bodies.

Arch. Rat. Mech. Analysis 41, pp. 108-120

1971b Formal first integrals of ordinary differential equations.

Rend. Circ. Mat. Palermo, Ser. II, Vol. 20, pp. 43-52

1973a Die topologische Struktur der Zeit.

Philosophia Naturalis

1973b Erweiterungen erster Integrale analytischer Differentialgleichungen.

Manuskript

Laugwitz, D.

1965 Differential and Riemannian Geometry.

Academic Press, New York-London

Lee, P. F.

1970 Periodic Orbits of Dynamical Systems.

Ph. D. Thesis, Georgia Institute of Technology, Atlanta, Ga.

Lefschetz, S.

1952 Differential Equations: Geometric Theory.

Interscience Publ., New York-London

Mach, E.

1883 Die Mechanik in ihrer Entwicklang.

English Translation: The Science of Mechanics.

LaSalle, Ill. 5th ed. 1942

MacLane, S.

1970 Hamiltonian Mechanics and Geometry.

Amer. Math. Monthly 77, pp. 520-586.

Milnor, J. W.

1965 Topology From the Differentiable Viewpoint.

U. Virginia Press

1969 Morse Theory.

Princeton U. P.

Munkres, J. R.

1966 Elementary Differential Topology.

Princeton U. P.

Nemytskii, V. V., and V. V. Stepanov

1960 Qualitative Theory of Differential Equations.

Princeton U. P.

Newton, I.

1686 Philosophiae Naturalis Principia Mathematica.

English translation 1729, revised 1962:

U. California Press

Poincaré, H.

1903 La science et l'hypothèse.

English translation: Dover, New York, 1952

Polites, G. W.

1968 An Introduction to the Theory of Groups.

International Textbook Co., Scranton, Pa.

Pontrjagin, L. S.

1957 Topologische Gruppen, Vol. I.

1958 Topologische Gruppen, Vol. II.

Teubner, Leipzig

Rinow, W.

1961 Die innere Geometric der metrischen Räume.

Springer, Berlin-Göttingen-Heidelberg

Rund, H.

1959 The Differential Geometry of Finsler Spaces.

Springer, Berlin-Göttingen-Heidelberg

Russell, B.

1903 The Principles of Mathematics.

8th ed. 1964. Allen & Unwin, London

Sansone, G., and R. Conti

1964 Non-linear Differential Equations.

Pergamon Press, Oxford and New York

Schubert, H.

1968 Topologie.

Teubner, Stuttgart

Schürer, M.

1943 Beitrag zur Dynamik der Sternsysteme.

Astron. Nachrichten 273

Shanmugadhasam, S.

1963 Generalized canonical formalism for degenerate dynamical systems.

Proc. Cambridge Philos. Soc. 59, pp. 743-757

Siegel, C. L.

1956 Vorlesungen uber Himmelsmechanik.

Springer, Berlin-Göttingen-Heidelberg

Siegel, C. L., and J. K. Moser

1971 Lectures on celestial mechanics.

Springer, Berlin-New York

Stoker, J. J.

1969 Differential Geometry.

Interscience Publ., New York

Suppes, P.

1967 Introduction to Logic.

Van Nostrand, Princeton

Szebehely, V.

1967 Theory of Orbits.

Academic Press, New York

Tricomi, F.

1961 Differential Equations.

Hafner, New York

Van der Waerden, B.

1964 Algebra, Vol. I.

Springer, Berlin-Gottingen-Heidelberg

Whittaker, E. T.

1961 A Treatise on the Analytical Dynamics of Particles and Rigid Bodies.

Cambridge U. P. (1st edition: 1904)

Wilson, H. K.

1971 Ordinary Differential Equations.

Addison-Wesley, Reading, Mass.

Wintner, A.

1952 The Analytical Foundations of Celestial Mechanics.

Princeton U. P.

SUPPLEMENTARY LITERATURE

Abraham, R.

1967 Foundations of Mechanics.

Benjamin, New York

Anslander, J., N. P. Bhatia, and P. Seibert

1964 Attractors in Dynamical Systems.

Bol. Soc. Mat. Mexicana, 1964, pp. 55-66

Bernfeld, S. R.

1972 Strongly Bounded Differential Systems.

Rend. Circ. Mat. Palermo, Sec. II Vol. 21, pp. 201-208

Bhatia, N. P.

On Asymptotic Stability in Dynamical Systems.

Math. Systems Theory 1, pp. 113-127

Stability and Liapunov Functions in Dynamical Systems.

Contr. Diff. Equations 3, pp. 175-188

1965 On Exponential Stability of Linear Differential Systems.

J. SIAM Control 2, pp. 181-191

1966 Some Oscillation Theorems for Second Order Differential Equations.

J. Math. Analysis and Applications 15, pp. 442-446

Bhatia, N. P., and G. P. Szegö

1967 An Extension Theorem for Asymptotic Stability and Its Application to a Control Problem.

In: Differential Equations and Dynamical Systems.

Academic Press, New York

1967 Stability Theory of Dynamical Systems.

Springer, New York

Epstein, S. T.

1964 General solutions of the Hamilton-Jacobi equation.

Amer. J. Phys. 32, pp. 688-690

Maharam, D.

1964 Incompressible transformations.

Fund. Math. 56, pp. 35-50

Mayne, G.

1964 Le problème de la séparation des variables pour les systèmes dynamiques scléronômes de fonction hamiltonienne quadratique mais non homogène en les momentoïdes.

C. r. Acad. Sci., Paris 258, pp. 61-63

Moser, J.

1968 Lectures on Hamiltonian Systems.

Amer. Math. Soc., Providence, R. I.

Pais, A. H. G.

1964 On the integrals of some types of Hamiltonian systems.

Univ. Lisbon, Revista Fac. Ci., II. Ser. A 10, pp. 131-38

Plummer, H. C.

1918 An Introductory Treatise on Dynamical Astronomy.

Cambridge U. P.

Pollard, H.

1966 Mathematical Introduction to Celestial Mechanics.

Prentice Hall, Englewood Cliffs, N. J.

Rüssmann, H.

1964 Über das Verhalten analytischer Hamiltonscher Differentialgleichungen in der Nähe einer Gleichgewichtslösung.

Math. Ann. 154, pp. 285-300

Ryll-Nardzewski, C.

1964 On Borel measurability of orbits.

Fund. Math. 56, pp. 129-130

Scott, D.

1964 Invariant Borel Sets.

Fund. Math. 56, pp. 117-128

Smale, S.

1967 Differentiable Dynamical Systems.

Bull. Amer. Math. Soc. 73, pp. 747-817

INDEX

The numbers indicate sections and subsections.

OTHER TITLES IN THE SERIES IN PURE AND APPLIED MATHEMATICS

Vol. 1. WALLACE—*An Introduction to Algebraic Topology*
Vol. 2. PEDOE—*Circles*
Vol. 3. SPAIN—*Analytical Conics*
Vol. 4. MIKHLIN—*Integral Equations*
Vol. 5. EGGLESTON—*Problems in Euclidean Space: Application of Convexity*
Vol. 6. WALLACE—*Homology Theory on Algebraic Varieties*
Vol. 7. NOBLE—*Methods Based on the Wiener–Hopf Technique for the Solution of Partial Differential Equations*
Vol. 8. MIKUSINSKI—*Operational Calculus*
Vol. 9. HEINE—*Group Theory in Quantum Mechanics*
Vol. 10. BLAND—*The Theory of Linear Viscoelasticity*
Vol. 11. KURTH—*Axiomatics of Classical Statistical Mechanics*
Vol. 12. FUCHS—*Abelian Groups*
Vol. 13. KURATOWSKI—*Introduction to Set Theory and Topology*
Vol. 14. SPAIN—*Analytical Quadrics*
Vol. 15. HARTMAN and MIKUSINSKI—*Theory of Lebesgue Measure and Integration*
Vol. 16. KULCZYCKI—*Non-Euclidean Geometry*
Vol. 17. KURATOWSKI—*Introduction to Calculus*
Vol. 18. GERONIMUS—*Polynomials Orthogonal on a Circle and Interval*
Vol. 19. ELSGOLC—*Calculus of Variations*
Vol. 20. ALEXITS—*Convergence Problems of Orthogonal Series*
Vol. 21. FUCHS and LEVIN—*Functions of a Complex Variable, Volume II*
Vol. 22. GOODSTEIN—*Fundamental Concepts of Mathematics*
Vol. 23. KEENE—*Abstract Sets and Finite Ordinals*
Vol. 24. DITKIN and PRUDNIKOV—*Operational Calculus in Two Variables and its Applications*
Vol. 25. VEKUA—*Generalized Analytic Functions*
Vol. 26. FASS and AMIR-MOÉZ—*Elements of Linear Spaces*
Vol. 27. GRADSHTEIN—*Direct and Converse Theorems*
Vol. 28. FUCHS—*Partially Ordered Algebraic Systems*
Vol. 29. POSTNIKOV—*Foundations of Galois Theory*
Vol. 30. BERMANT—*A Course of Mathematical Analysis, Part II*
Vol. 31. LUKASIEWICZ—*Elements of Mathematical Logic*
Vol. 32. VULIKH—*Introduction to Functional Analysis for Scientists and Technologists*
Vol. 33. PEDOE—*An Introduction to Projective Geometry*
Vol. 34. TIMAN—*Theory of Approximation of Functions of a Real Variable*
Vol. 35. CSÁSZÁR—*Foundations of General Topology*
Vol. 36. BRONSHTEIN and SEMENDYAYEV—*A Guide-Book to Mathematics for Technologists and Engineers*
Vol. 37. MOSTOWSKI and STARK—*Introduction to Higher Algebra*
Vol. 38. GODDARD—*Mathematical Techniques of Operational Research*
Vol. 39. TIKHONOV and SAMARSKII—*Equations of Mathematical Physics*
Vol. 40. MCLEOD—*Introduction to Fluid Dynamics*
Vol. 41. MOISIL—*The Algebraic Theory of Switching Circuits*
Vol. 42. OTTO—*Nomography*
Vol. 43. RANKIN—*An Introduction to Mathematical Analysis*
Vol. 44. BERMANT—*A Course of Mathematical Analysis, Part I*
Vol. 45. KRASNOSEL'SKII—*Topological Methods in the Theory of Nonlinear Integral Equations*
Vol. 46. KANTOROVICH and AKHILOV—*Functional Analysis in Normed Spaces*
Vol. 47. JONES—*The Theory of Electromagnetism*

Vol. 48. FEJES TÓTH—*Regular Figures*
Vol. 49. YANO—*Differential Geometry on Complex and Almost Complex Spaces*
Vol. 50. MIKHLIN—*Variational Methods in Mathematical Physics*
Vol. 51. FUCHS and SHABAT—*Functions of a Complex Variable and Some of their Applications, Volume I*
Vol. 52. BUDAK, SAMARSKII and TIKHONOV—*A Collection of Problems on Mathematical Physics*
Vol. 53. GILES—*Mathematical Foundations of Thermodynamics*
Vol. 54. SAUL'YEV—*Integration of Equations of Parabolic Type by the Method of Nets*
Vol. 55. PONTRYAGIN *et al.—The Mathematical Theory of Optimal Processes*
Vol. 56. SOBOLEV—*Partial Differential Equations of Mathematical Physics*
Vol. 57. SMIRNOV—*A Course of Higher Mathematics, Volume I*
Vol. 58. SMIRNOV—*A Course of Higher Mathematics, Volume II*
Vol. 59. SMIRNOV—*A Course of Higher Mathematics, Volume III, Part 1*
Vol. 60. SMIRNOV—*A Course of Higher Mathematics, Volume III, Part 2*
Vol. 61. SMIRNOV—*A Course of Higher Mathematics, Volume IV*
Vol. 62. SMIRNOV—*A Course of Higher Mathematics, Volume V*
Vol. 63. NAIMARK—*Linear Representations of the Lorentz Group*
Vol. 64. BERMAN—*A Collection of Problems on a Course of Mathematical Analysis*
Vol. 65. MESHCHERSKII—*A Collection of Problems of Mechanics*
Vol. 66. ASCOTT—*Periodic Differential Equations*
Vol. 67. SANSONE and CONTI—*Non-linear Differential Equations*
Vol. 68. VOLKOVYSKII, LUNTS and ARAMANOVICH—*A Collection of Problems on Complex, Analysis*
Vol. 69. LYUSTERNIK and YANPOL'SKII—*Mathematical Analysis—Functions, Limits, Series, Continued Fractions*
Vol. 70. KUROSH—*Lectures in General Algebra*
Vol. 71. BASTON—*Some Properties of Polyhedra in Euclidean Space*
Vol. 72. FIKHTENGOL'TS—*The Fundamentals of Mathematical Analysis, Volume 1*
Vol. 73. FIKHTENGOL'TS—*The Fundamentals of Mathematical Analysis, Volume 2*
Vol. 74. PREISENDORFER—*Radiative Transfer on Discrete Spaces*
Vol. 75. FADDEYEV and SOMINSKII—*Elementary Algebra*
Vol. 76. LYUSTERNIK, CHERVONENKIS and YANPOL'SKII—*Handbook for Computing Elementary Functions*
Vol. 77. SHILOV—*Mathematical Analysis—A Special Course*
Vol. 78. DITKIN and PRUDNIKOV—*Integral Transforms and Operational Calculus*
Vol. 79. POLOZHII—*The Method of Summary Representation for Numerical Solution of Problems of Mathematical Physics*
Vol. 80. MISHINA and PROSKURYAKOV—*Higher Algebra—Linear Algebra, Polynomials, General Algebra*
Vol. 81. ARAMANOVICH *et al.—Mathematical Analysis—Differentiation and Integration*
Vol. 82. REDEI—*The Theory of Finitely Generated Commutative Semigroups*
Vol. 83. MIKHLIN—*Multidimensional Singular Integrals and Integral Equations*
Vol. 84. LEBEDEV, SKAL'SKAYA and UFLYAND—*Problems in Mathematical Physics*
Vol. 85. GAKHOV—*Boundary Value Problems*
Vol. 86. PHILLIPS—*Some Topics in Complex Analysis*
Vol. 87. SHREIDER—*The Monte Carlo Method*
Vol. 88. POGORZELSKI—*Integral Equations and their Applications, Vol. I, Parts 1, 2 and 3*
Vol. 89. SVESHNIKOV—*Applied Methods of the Theory of Random Functions*
Vol. 90. GUTER, KUDRYAVTSEV and LEVITAN—*Elements of the Theory of Functions*
Vol. 91. REDEI—*Algebra, Vol. I*
Vol. 92. GEL'FOND and LINNIK—*Elementary Methods in the Analytic Theory of Numbers*
Vol. 93. GUREVICH—*The Theory of Jets in an Ideal Fluid*
Vol. 94. LANCASTER—*Lambda-matrices and Vibrating Systems*
Vol. 95. DINCULEANU—*Vector Measures*
Vol. 96. SLUPECKI and BORKOWSKI—*Elements of Mathematical Logic and Set Theory*
Vol. 97. REDEI—*Foundations of Euclidean and Non-Euclidean Geometries according to F. Klein*
Vol. 98. MACROBERT—*Spherical Harmonics*
Vol. 99. KUIPERS-TIMMAN—*Handbook of Mathematics*
Vol. 100. SALOMAA—*Theory of Automata*
Vol. 101. KURATOWSKI—*Introduction to Set Theory and Topology* (2nd Edition)
Vol. 102. BLYTH and JANOWITZ—*Residuation Theory*
Vol. 103. KOSTEN—*Stochastic Theory of Service Systems*
Vol. 104. WAN—*Lie Algebras*